Plasma Gasification and Pyrolysis

Currently, the most widely used treatment of waste is thermal processing, such as incineration. However, thermal plasma technologies offer alternative, cutting-edge, and environmentally friendly processes, which are also considered more energy-efficient and safe.

This book provides a comprehensive yet accessible introduction to the process of thermal plasma gasification and pyrolysis.

It is an ideal guide for graduate students pursuing further studies in plasma technologies and engineering, in addition to early-career researchers and scientists from related areas looking for material contextual to their own subject matter.

Features

- Presents an interdisciplinary approach, applicable to a wide range of researchers in waste treatment companies, authorities, and energy and environmental policymakers.
- Authored by authorities in the field.
- Up to date with the latest developments and technologies.

Plasma Gasification and Pyrolysis

Milan Hrabovsky, Michal Jeremiáš and
Guido van Oost

CRC Press
Taylor & Francis Group
Boca Raton London New York

CRC Press is an imprint of the
Taylor & Francis Group, an **informa** business

First edition published 2023
by CRC Press
6000 Broken Sound Parkway NW, Suite 300, Boca Raton, FL 33487-2742

and by CRC Press
4 Park Square, Milton Park, Abingdon, Oxon, OX14 4RN
CRC Press is an imprint of Taylor & Francis Group, LLC

© 2023 Milan Hrabovsky, Michal Jeremiáš and Guido Van Oost

Reasonable efforts have been made to publish reliable data and information, but the author and publisher cannot assume responsibility for the validity of all materials or the consequences of their use. The authors and publishers have attempted to trace the copyright holders of all material reproduced in this publication and apologize to copyright holders if permission to publish in this form has not been obtained. If any copyright material has not been acknowledged please write and let us know so we may rectify in any future reprint.

Except as permitted under U.S. Copyright Law, no part of this book may be reprinted, reproduced, transmitted, or utilized in any form by any electronic, mechanical, or other means, now known or hereafter invented, including photocopying, microfilming, and recording, or in any information storage or retrieval system, without written permission from the publishers.

For permission to photocopy or use material electronically from this work, access www.copyright.com or contact the Copyright Clearance Center, Inc. (CCC), 222 Rosewood Drive, Danvers, MA 01923, 978-750-8400. For works that are not available on CCC please contact mpkbookspermissions@tandf.co.uk

Trademark notice: Product or corporate names may be trademarks or registered trademarks and are used only for identification and explanation without intent to infringe.

ISBN: 978-0-367-55685-3 (hbk)
ISBN: 978-0-367-56230-4 (pbk)
ISBN: 978-1-003-09688-7 (ebk)

DOI: 10.1201/9781003096887

Typeset in Minion
by Deanta Global Publishing Services, Chennai, India

Contents

Preface, ix

About the Authors, xi

CHAPTER 1 ▪ Basic Features of Plasmas 1
GUIDO VAN OOST
1.1 DEFINITION OF A PLASMA 1
1.2 THE TEMPERATURE OF A PLASMA (IN eV) 1
1.3 PLASMA, THE 'FOURTH STATE OF MATTER' 3
1.4 COMPARISON OF THE STATES OF MATTER 5
1.5 ELECTRICAL NEUTRALITY OF A PLASMA 6
1.6 CLASSIFICATION OF PLASMAS 7
1.7 MATHEMATICAL DESCRIPTIONS OF PLASMAS 11
1.8 THERMAL VERSUS NON-THERMAL PLASMAS 12
1.9 METHODS OF PLASMA GENERATION 12
REFERENCE 13

CHAPTER 2 ▪ Generation of Thermal Plasmas 15
MILAN HRABOVSKÝ
2.1 TYPES OF ARC PLASMA GENERATORS 16
2.2 PROCESSES AND PROPERTIES OF ARC DISCHARGE 20
    2.2.1 Cathode Region 21
    2.2.2 Anode Region 23
    2.2.3 Arc Column 24

2.3 EFFECT OF PLASMA GAS PROPERTIES AND ARC CHAMBER DIMENSIONS ON TORCH CHARACTERISTICS 26
REFERENCES 31

CHAPTER 3 ▪ Conventional Thermochemical Technologies for Waste Treatment 33

MICHAL JEREMIÁŠ

3.1 WASTE INCINERATION 34
3.2 LARGE-SCALE WASTE GASIFICATION 36
3.3 USE OF GASIFICATION TECHNOLOGY FOR WASTE TREATMENT ON A SMALLER SCALE 37
3.4 FURTHER POSSIBLE DEVELOPMENT OF TECHNOLOGIES USING WASTE GASIFICATION AND PYROLYSIS 38
REFERENCES 40

CHAPTER 4 ▪ Thermal Plasma Waste Treatment 45

MILAN HRABOVSKÝ

4.1 CHARACTERISTICS OF PLASMA PROCESSING 45
4.2 PROCESSES IN PLASMA REACTOR 47
4.3 PLASMA PYROLYSIS, GASIFICATION, AND PLASMA-AIDED COMBUSTION 51
4.4 PLASMA-MATERIAL HEAT TRANSFER AND PROCESS RATE 56
4.5 ENERGY BALANCE OF PLASMA TREATMENT 62
4.6 GASIFICATION OF VARIOUS ORGANICS IN STEAM PLASMA 70
REFERENCES 77

CHAPTER 5 ▪ Product Applications 81

GUIDO VAN OOST

5.1 INTRODUCTION 81
5.2 SINGLE VERSUS TWO-STAGE PLASMA GASIFICATION 84

5.3	APPLICATIONS		85
	5.3.1	Municipal Solid Waste	85
	5.3.2	Biomass	87
	5.3.3	Hazardous Materials	88
	5.3.4	Sludge	89
5.4	EXAMPLES OF PLASMA GASIFICATION PLANTS AT AN INDUSTRIAL SCALE		89
	5.4.1	Westinghouse Plasma Corporation	92
	5.4.2	Europlasma	94
	5.4.3	Tetronics	94
	5.4.4	Phoenix Solutions	94
5.5	ECONOMICAL ASPECTS		95
5.6	ENVIRONMENTAL ASPECTS		98
REFERENCES			99

INDEX, 101

Preface

THE MOST WIDELY USED treatment of waste is thermal processing such as incineration. The alternative environmentally friendly process treated in this book is based on thermal plasma technology which is a very flexible tool because it allows operating in a wide temperature range with almost any chemical composition of waste and chemicals needed for processing this waste. It allows the conversion of organic waste into energy or chemical substances as well as the destruction of toxic organic compounds in a scenario that for each specific type of waste can be considered optimal, in terms of both energy efficiency and environmental safety. The increasingly stringent legislation on the treatment of waste streams and the limitations of conventional technologies render plasma technologies more and more attractive.

The driving force behind the book is to give priority to environmental quality at an affordable cost and to use innovative thermochemical conversion (gasification and pyrolysis) technologies to contribute to sustainable development and circular economy in which waste is managed as a resource.

The book is intended as a work of tertiary literature. As such, it contains digested knowledge in an easily accessible format. It contains neither research literature (primary literature) nor review articles summarising original papers (secondary literature). The content therefore consists of established information in this field. The level of contributions is such that a master's student can benefit from contributions relevant for the specific area of his/her research and at a later stage from contributions that are not in his or her area of expertise. Contributions will start at a level that is appropriate and accessible for all early-career researchers and scientists from related areas looking for material contextual to their own subject matter and should develop to the most appropriate level of advanced research.

About the Authors

Guido van Oost is emeritus professor of Nuclear Fusion at Ghent University, Belgium. He is Honorary Professor of the Peter the Great St. Petersburg Polytechnic University, and holds a Silver Commemorative Medal of Charles University in Prague, an Ernst Mach Honorary Medal for Merit in the Physical Sciences (Czech Academy of Sciences), and a Medal of the Czech Technical University in Prague to honour outstanding contributions to the development of the Faculty of Nuclear Sciences and Physical Engineering.

Milan Hrabovsky is a senior scientist and consultant at Institute of Plasma Physics, Academy of Sciences of the Czech Republic. From 1967 to 1990 he worked as a researcher and later as the Head of Department of Switchgear Technology at the Institute of Electrical Engineering, Prague. His main research areas were vacuum breakdown processes, physics of electric arcs in vacuum and gases, basic problems of current interruption and electromagnetic plasma launchers. In 1990 he joined the Institute of Plasma Physics, Academy of Sciences of the Czech Republic. He founded the Department of Thermal Plasmas and he has worked as a head of the Department till his retirement in 2018.

Michal Jeremiáš is head of department at the Institute of Plasma Physics of the Czech Academy of Science (CAS). He was previously researcher at the Institute of Chemical Process Fundamentals at CAS. He is the Principal Investigator at two national projects, a member of the team of four international projects and two UK projects.

CHAPTER 1

Basic Features of Plasmas

Guido Van Oost

1.1 DEFINITION OF A PLASMA

A plasma is a quasi-neutral ensemble of positively and negatively charged particles and neutral particles, featuring a *collective behaviour* because local charge separation and electrical currents create additional electric and magnetic fields which together with externally imposed fields determine the equations of motion of the individual particles.

In principle, this ensemble of particles is fully determined by the laws of classical mechanics, the Lorentz force, and the Maxwell equations.

1.2 THE TEMPERATURE OF A PLASMA (IN eV)

Density and temperature are the two most important parameters characterising a plasma. Consider particles that mutually exchange energy through multiple elastic collisions, and assume that this collection of particles is in *thermodynamic equilibrium* at temperature T. An energy distribution will be established given by a characteristic function which in the simplest case is the one-dimensional *Maxwell distribution*.

This Maxwell velocity distribution is given by

$$f(v)dv = \left(\frac{m}{2\pi kT}\right)^{\frac{3}{2}} 4\pi v^2 \exp\left(-\frac{mv^2}{2kT}\right) dv, \qquad (1.1)$$

DOI: 10.1201/9781003096887-1

where v is the particle velocity, and k is the Boltzmann constant. The function $f(v)$ gives the probability that a particle has a velocity between v and $v + dv$. Since all particles have a velocity between 0 and ∞,

$$\int_0^\infty f(v)dv = 1,$$

in other words, the velocity distribution function is normalised to 1.

In Equation (1.1), the temperature T is an important parameter (see Figure 1.1). The 'most probable velocity' v_w occurs at the maximum of $f(v)$ and can be found by putting $df/dv = 0$, i.e.,

$$v_w = \left(\frac{2kT}{m}\right)^{\frac{1}{2}}. \tag{1.2}$$

Herewith the most probable energy ε_w corresponds:

$$\varepsilon_w = \frac{1}{2}mv_w^2 = kT. \tag{1.3}$$

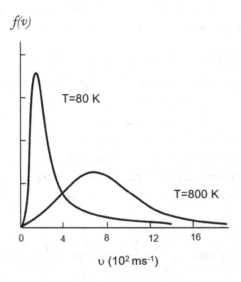

FIGURE 1.1 The function $f(v)$ gives the probability that a particle has a velocity between v and $v + dv$.

In plasma physics, relation (1.3) is generally used to express plasma temperatures in eV (direct relation between energy and temperature). For example, an energy of $\varepsilon = 1\,\text{eV}$ corresponds to a temperature given by

$$1\,\text{eV} = 1.602\,10^{-19}\,\text{J} = kT,$$

or (with $k = 1.38\,10^{-23}$ J/K = Boltzmann constant):

$$T = 1.602\,10^{-19}/1.38\,10^{-23} = 1.16\,10^{4}\,\text{K}.$$

Hence,

$$1\,\text{eV} \Leftrightarrow 11{,}605\,\text{K}, \tag{1.4}$$

(rule of thumb: 1 eV ⇔ 10,000 K). For example, room temperature (300 K) thus corresponds to 0.025 eV; 1 keV ⇔ 10 million K; 10 keV ⇔ 100 million K (nuclear fusion).

From Equation (1.1) follows that the *average kinetic energy* $<\varepsilon>$ of the particles

$$\langle\varepsilon\rangle = \left\langle\frac{1}{2}mv^2\right\rangle = \frac{3}{2}kT, \tag{1.5}$$

where $\langle v^2 \rangle$ the average velocity squared is given by

$$\langle v^2 \rangle = \int_0^\infty v^2 f(v)\,dv,$$

From Equations (1.3) and (1.5), it follows that a plasma with temperature 1 keV has an average kinetic energy of 1.5 keV.

1.3 PLASMA, THE 'FOURTH STATE OF MATTER'

Plasma as the fourth state of matter was introduced by Crookes in 1879. The name plasma was used for the first time by Langmuir in the 1920s for the description of the positive column of a gas discharge. Prior to roughly 1940, plasma physics was a rather limited and specialised branch of physics, with applications primarily in the field of light technology. However, around 1940, it became clear that the interaction between electrically charged matter and electromagnetic fields constitutes a vast research domain with many applications in geophysics, stellar and cosmic physics, space research, and plasma chemistry and in the technical utilisation of gas discharges. The most ambitious plasma physics project is the development of *thermonuclear fusion* as the energy source for the future (fusion physics).

We specifically mention Hannes Alfvén (Stockholm 1908–1995), the winner of the 1970 Nobel Prize in Physics and founder (1942) of plasma physics, particularly for his work on magnetohydrodynamics (MHD; the effects of magnetic and electric fields).

Plasma physics in fact describes the 'fourth state of matter,' covering practically the whole domain of classical physics, as well as important parts of atomic and nuclear physics. By its relationship with different branches of physics, plasma physics has always stimulated a strong interaction between fundamental research and technological developments.

There exists an obvious analogy between the four states of matter and the four Greek elements earth, water, air, and fire (plasma). The word plasma is deduced from an ancient Greek word meaning mouldable substance, without definite shape. A common point is that the plasma easily escapes from our control, just like a moulded work glides between your fingers. The association with biological plasmas is totally wrong, although the researcher often gets the impression that the obstinate plasma lives its own life.

More than 90% of the known matter consists of plasma. The sun and the stars are gigantic masses of hot plasma. The sun consists of 92% of hydrogen plasma and 7.8% of helium plasma (0.2% of others); its interior temperatures exceed 10^7 K. In the universe, plasmas occur much more frequently than the three other states of matter. However, on earth, plasma is rather exceptional.

Plasmas are in the first place interesting because of their variety of properties. Plasmas have more and more important and intriguing applications.

The characteristics of plasma are the occurrence of charged particles: electrons, ions (positive as well as negative), and possibly charged clusters of dust particles ('dusty plasmas'), besides neutral particles (atoms, molecules, radicals). Due to the occurrence of charged particles, plasma can conduct electrical current, can be affected by electric and magnetic fields, and can be excited and hence emit radiation (applied in lighting and gas lasers). This requires a high temperature of a minimum of a few tens of thousands of degrees (or a few eV) and is the reason why plasma occurs only sporadically on earth.

This high temperature implies that plasma is chemically very active (fast chemical conversion or modification with relatively few particles). This is the basis for many recent applications such as plasma deposition and etching, surface modification, environmentally friendly treatment of waste streams with plasma torches, and so on. Nuclear fusion research delivered many new insights into the medium plasma and is a very important incentive for plasma research.

1.4 COMPARISON OF THE STATES OF MATTER

The states of matter are characterised by their average energy per particle (atom or molecule). The solid matter has the lowest energy, while plasma has the highest. The order of magnitude of the energy per particle is illustrated in Table 1.1, taking an example of water. The transition energies are given in eV per molecule.

If a gas is further heated, the molecules will dissociate. If the thermal motion is strong enough, then more and more atoms will be split into electrons and ions: the gas becomes ionised and transformed into a plasma. The degree of ionisation is determined by the electron temperature relative to the ionisation energy (and more weakly by the density), in a relationship called the Saha ionisation equation.

In contrast with the other phase transitions, ionisation does not occur at a precise temperature (at a given pressure). The transition of the non-ionised to the ionised state happens gradually over a large temperature

TABLE 1.1 Comparison of Transition Temperatures and Transition Energies (in the Case of Water)

	T (K) ↑	
Plasma		
Ionisation	10^4–10^5 K	Ionisation energy $H \rightarrow H^+ + e^-$: 13.6 eV/atom
		$O \rightarrow O^+ + e^-$: 12.5 eV/atom
Gas		Dissociation energy: about 5 eV
Boiling point	373 K	Evaporation energy: 0.42 eV/molecule
Liquid		0–100°C: 0.08 eV/molecule
Melting point	273 K	Melting energy: 0.06 eV/molecule
Solid		

interval. Furthermore, besides thermal ionisation, there occurs also ionisation, e.g., by electron impact or short-wavelength electromagnetic radiation. Therefore, from the thermodynamic point of view, the term 'fourth state of matter' is not fully suitable. On the other hand, due to the presence of charged particles, the characteristics of plasmas are so different from those of non-ionised gases that plasmas indeed merit the status of the state of matter.

1.5 ELECTRICAL NEUTRALITY OF A PLASMA

The plasma is macroscopically quasi-neutral, but due to thermal motion, the electrical neutrality is temporarily or locally perturbed. When local deviations of electrical neutrality occur in the plasma, then electrostatic fields are established that try to restore electrical neutrality through an influx and/or outflux of suitable neutralising charges. Due to Coulombic forces, the charged particles locally oscillate around an average equilibrium state. Note that in the absence of external energy sources, the only possibility to build up electric field energy is by thermal motion (this is the case treated here).

Electrical neutrality is only *macroscopically* satisfied, i.e., by averaging

- Over a 'sufficiently large' volume, at a given instant t.
- During a 'sufficiently long' time, at a given place.

Microscopically, electrical neutrality is not satisfied in the following conditions:

- If at a given instant t (snap shot), a too small plasma volume is studied. The volume within which electrical neutrality is not satisfied has a characteristic dimension, the *Debye length* λ_D (critical dimension for collective behaviour).
- If at a given place, the plasma is observed during a too short time interval. The time interval within which electrical neutrality is not satisfied has a characteristic duration: the plasma period τ_P (*plasma frequency*).

The Debye length or screening length is a rather universal concept. The Dutchman Peter Debye introduced this quantity for the study of electrolytes (1923). Irving Langmuir took it over in plasma physics (1928), and not just as one of many other concepts: he gave the definition of plasma in terms of the Debye length.

1.6 CLASSIFICATION OF PLASMAS

Density and temperature are the two most important parameters characterising a plasma. Consider a plasma in thermal equilibrium. The particles have an average thermal energy of

$$\langle \varepsilon \rangle = \frac{1}{2} m \langle v \rangle^2 = \frac{3}{2} kT.$$

The average distance d between the particles fulfils the condition

$$nd^3 = 1 \quad \text{or} \quad d = n^{-\frac{1}{3}}.$$

Other characteristic lengths are the Debye length λ_D and the de Broglie wavelength $\lambda_{d_B} = \dfrac{h}{mv}$.

The classification of plasmas is represented in a log n versus log T diagram. The domain of the existence of plasmas is much broader than for the three other states of matter. Several zones can be distinguished.

(a) **Relativistic plasma**

It is a plasma with such high electron energy that the classical (non-relativistic) mechanics is no longer valid. In such a case (thermal energy > rest mass electron) (Figure 1.2):

$$kT > m_{0,e} c^2.$$

In the diagram, the boundary line is drawn for which $kT = m_{0,e} c^2$.

With $m_{0,e} = 9.1\,10^{-31}$ kg, this gives numerically

$$\log T = 9.77. \tag{1.6}$$

(b) **Quantum plasma or degenerate plasma**

It is a plasma with such high particle density that the classical Maxwell–Boltzmann statistics can no longer be applied since

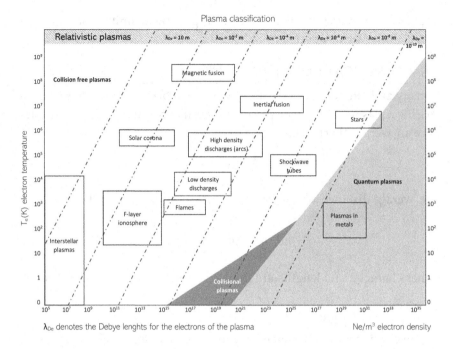

FIGURE 1.2 The classification of plasmas is represented in a log n vs. log T diagram.

quantum effects are dominating. Quantum effects dominate when the de Broglie wavelength is larger than d, thus

$$\frac{h}{mv} > d. \quad (1.7)$$

The particle density is so high (i.e., d so small) that a particle can no longer be described as a separate individual, but as a particle that interferes quantum-mechanically with many other particles (through its wave function). The plasma behaves as a 'Fermi–Dirac liquid.' Expressing Equation (1.7) as a function of n and T, one finds

$$\frac{h}{\sqrt{3mkT}} > n^{\frac{-1}{3}} \quad \text{or} \quad n > \left(\frac{3mkT}{h^2}\right)^{\frac{3}{2}}. \quad (1.8)$$

In the diagram, the following boundary line is drawn:

$$n = \left(\frac{3mkT}{h^2}\right)^{\frac{3}{2}}$$

or

$$\log T = \left(\frac{2}{3}\right)\log n + \log\left(\frac{h^2}{3mk}\right). \quad (1.9)$$

Numerically:

$$\log T = \left(\frac{2}{3}\right)\log n - 13.93.$$

Condition (1.8) can also be written as follows:

$$kT < \frac{h^2}{3m} n^{\frac{2}{3}}.$$

Taking into account the formula for the energy of the Fermi level of the electron gas in a metal

$$E_f = \frac{h^2}{2m}\left(\frac{3n}{8\pi}\right)^{\frac{2}{3}},$$

this gives

$$kT < \frac{2}{3}\left(\frac{8\pi}{3}\right)^{\frac{2}{3}} E_f = 2{,}75\, E_f.$$

In a quantum plasma, the average kinetic energy is small with respect to the energy of the Fermi level.

(c) **Collisional plasma**

Also named 'collision-dominated plasma' (strongly coupled plasma), it is a plasma where the motion of the particles is not due to the interaction with many other particles but results from rather random individual interactions ('binary collisions') with other particles. The condition under which binary collisions dominate thermal effects is

$$\lambda_D < d \quad (1.10)$$

or $\lambda_D^3 \, n < 1$, i.e., the Debye sphere contains practically no particles. In the diagram, the following boundary line is drawn

$$\lambda_D < d. \tag{1.11}$$

Expressed as a function of n and T, and considering $\lambda_D = 69\sqrt{\dfrac{T}{n}}$, this gives

$$\log T = +\frac{1}{3}\log n - 5.68.$$

(d) **Classical plasma**

It is, according to Langmuir (1928), a quasi-neutral gas with charged and neutral particles which exhibits a *collective behaviour*.

Conditions:

$$d \ll \lambda_D < L,$$

where L is a characteristic dimension of the gas container.

The condition $\lambda_D < L$ guarantees *quasi-neutrality*. Only at distances closer than a few Debye lengths from the wall deviation from quasi-neutrality occurs.

The condition $d \ll \lambda_D$ guarantees the *collective behaviour* of the plasma. This condition is equivalent to

$$n\,\lambda_D^3 \gg 1 \quad \text{or} \quad \frac{4}{3}\pi\lambda_D^3\, n \gg 1,$$

i.e., the Debye sphere contains a large number of particles. The charged particles exert interactive Coulombic forces on all other charged particles of the plasma, even at large distance. Collective behaviour means that the particle motion is not determined by accidental local circumstances of a binary collision (short range interaction), but in the first place by Coulombic interactions (long range) with very many particles of the plasma; all particles within the Debye sphere contribute collectively. In such a 'collision-free' plasma ('collisionless plasma'), the forces occurring in local binary collisions can be neglected with respect to the electromagnetic forces.

In the log n – log T diagram, the lines of constant Debye length are drawn:

$$\lambda_D = 69\sqrt{\frac{T}{n}} = C^{te}.$$

These are lines given by the equation

$$\log T = \log n - 2\log 69 + 2\log C^{te},$$

or numerically:

$$\log T = \log n + C^{te}.$$

1.7 MATHEMATICAL DESCRIPTIONS OF PLASMAS

Some plasma properties are preferably described by studying the motion of *individual charged particles* in fields (electric, magnetic, potential fields, gravity field, etc.); this is a description at individual molecular level, i.e., from the *microscopic point of view*). Other properties can be better described as *collective ones*, whereby a plasma is considered as a collection of a very large number of charged particles, and statistical methods are applied: the plasma properties are considered as *statistically averaged properties*: description at a *collective molecular level*. This is a description from a *macroscopic point of view, the so-called fluid description*.

Furthermore, the plasma particles are *electrically charged*. The interaction between the particles is dominated by electric and magnetic forces instead of by the much weaker intermolecular (e.g., van der Waals) forces. As a result, the motions of these charged particles are extra coupled through $\bar{E}$ and $\bar{B}$ fields, and in a plasma collective phenomena occurrence which are unknown in a non-ionised gas. Hence, quite rightly, plasma is called the fourth state of matter. As a consequence, in a plasma, the following equations have to be fulfilled simultaneously and in a self-consistent way:

- The (macroscopic) hydromagnetic equations (HM) equations, also called MHD equations (magnetohydrodynamic equations).
- The equations of Maxwell.

1.8 THERMAL VERSUS NON-THERMAL PLASMAS

For the processing of materials and waste, non-thermal as well as thermal plasmas are utilised.

Thermal plasmas (hot plasmas) are characterised by their high energy densities and by the equal temperatures of the electrons and the heavy particles, i.e., thermal plasmas are in local thermodynamic equilibrium (LTE). Non-thermal plasmas ('cold' plasmas), on the other hand, are non-equilibrium ionised gases, which are characterised by lower energy densities and by the large difference between the electron temperatures and the heavy particle temperatures. Due to the temperature of ions and neutral particles being relatively low, cold plasmas can be used for low-temperature plasma chemistry and for the treatment of heat-sensitive materials including polymers and biological tissues.

The state of equilibrium of a plasma depends on the collision frequency and the energy exchange during a collision, which strongly depend on pressure. A high gas pressure implies many collisions in the plasma (i.e., a short collision means free path, compared to the discharge length), leading to an efficient energy exchange between the plasma species, and hence equal temperatures. A low gas pressure results in only a few collisions in the plasma (i.e., a long collision means free path compared to the discharge length), and hence different temperatures of the plasma species due to inefficient energy transfer.

This book deals with thermal plasmas. They are characterised by a high energy density and a high energy transfer rate, short reaction times for chemical reactions in the plasma, and a wide choice of plasma media. These characteristics make thermal plasmas suitable for a diversity of industrial applications and offer many research possibilities. Examples of industrial applications are cutting, welding, spraying, analysis by inductively coupled plasma, furnaces for metallurgy with DC arcs and graphite electrodes, tundish heating, metal melting and purification, and environmentally friendly treatment of waste streams with plasma torches (the topic of this book).

1.9 METHODS OF PLASMA GENERATION

Plasmas are formed by supplying energy to a neutral gas causing the formation of charge carriers. This can for example be achieved by driving an electric current through the gas, subjecting the gas to electromagnetic radiation, by supplying thermal energy, for example, in flames (where exothermic chemical reactions of the molecules are used as the prime energy source) or by applying adiabatic compression to the gas heating it up to the point of plasma generation.

The most common way for transferring energy to the working gas and generating plasma is by means of an electric field. At room temperature, gases consist of neutral species and are good insulators. To generate enough charge carriers and make the gas electrically conducting, a sufficiently high potential difference is applied between two electrodes placed in a gas. In a normal gas, the negatively charged electrons around the positive charge in the nucleus form an electrically neutral system. However, some stray electrons are always present in the gap between the electrodes, by cosmic rays or any other background radiation, photoelectric effect at the electrode surface due to the absorption of UV photons, or as a consequence of field emission from rough surfaces. These stray electrons are accelerated in the electric field and collide with gas molecules. If these collisions are inelastic, new electrons and ions are produced in the gas phase by different ionisation processes, leading to an avalanche of charged particles and a conductive path for an electric arc to form between the cathode and anode (electrical breakdown). Due to the electrical resistivity across the system significant heat is generated by the arc, which is first captured by the electrons because of their high mobility and strips them away from the gas molecules. The electrons transfer part of this absorbed energy to the heavy particles through elastic collisions.

In thermal plasmas with their high electron density and elastic collision frequencies, this energy transfer is important and leads to thermodynamic equilibrium. Arc discharges create a high density and high temperature region between the electrodes. With the aid of a sufficiently high gas flow in the electrode gap, the plasma extends beyond one of the electrodes (plasma jet), thereby transporting the plasma energy to the reaction region. A plasma torch, also known as plasmatron, is a device that generates a directed flow of thermal plasma from its nozzle. The primary electricity source of plasma torches can be direct current (DC), alternating current (AC) at the mains frequency (50 Hz) or AC at radiofrequency. Other characteristics of plasma torches are the arc stabilisation mechanism, the plasma gas, the types of flow, and the geometry and cooling of the electrodes (see Chapter 2).

REFERENCE

Langmuir, I., and Jones, H. A. (1928). "Collisions between electrons and gas molecules." *Physical Review* 31 (3): 357–404. doi:10.1103/PhysRev.31.357.

CHAPTER 2

Generation of Thermal Plasmas

Milan Hrabovský

THERMAL PLASMA TECHNOLOGIES ARE used in diverse industrial and scientific applications. These technologies include the production of new materials with unusual or superior properties, modification of materials and surfaces, treatment of waste materials, and gasification of organic substances. Thermal plasmas (Boulos et al. 1994) exhibit unique advantages such as high temperatures in the range of 5,000–50,000 K, high energy density, high energy transfer rate, and extremely low reaction times for chemical reactions.

A number of different types of plasma units are utilised in various plasma processing technologies. Thermal plasmas used in waste treatment and material gasification systems are commonly produced in electric arc discharges. Alternative types of discharges utilised in these technologies are inductively coupled radio frequency (RF) discharges and microwave plasma (MW).

Most industrial and laboratory systems are based on arc plasma torches. Plasma systems with power ranging from several kilowatts to tens of megawatts can be realized with arc plasma torches. The main problems of arc systems are erosion of electrodes and nozzles leading to limitations of time due to continuous operation of systems and reduction of lifetime of arc torches. In recent years, these problems led to experimenting with plasma generators based on inductively coupled RF discharges or microwave discharges. The

DOI: 10.1201/9781003096887-2

lifetime of these systems, which operate without electrodes, is higher than that of arc torches. However, the complexity and cost of power sources for RF or MW systems are the main limitations for the preparation of high-power RF or MW plasma systems. A limited number of RF and MW systems were realized with power not higher than 100 kW.

2.1 TYPES OF ARC PLASMA GENERATORS

The devices designed for the production of thermal plasma in arc discharges are called plasma torches, plasma guns, or plasmatrons. An electric arc in plasma torches is ignited between two electrodes which are incorporated into the body of the plasma torch (non-transferred arc), or one of the electrodes is positioned outside of the torch body (transferred arc). The outside electrode, usually an anode, is made of electrically conducting material, often a treated material. The discharge is ignited by electrical breakdown after application of high voltage between electrodes. A pulsed high-frequency voltage is commonly used for arc ignition in plasma torches. The discharge could also be started by touching and consequently separating the electrodes or by explosion of a wire connected between electrodes.

Arc discharges are inherently unstable. The principal goal of the design of arc plasma torches is stabilisation of the position of the discharge and control of plasma properties needed for a specific application. As the arc discharges have in most cases negative volt-ampere characteristics, the current in the electric circuit must be controlled either by elements of the circuits (e.g., by inductance in alternating current [AC] part of the circuit) or by using a power supply with a controlled current value. Both the direct current (DC) and AC arcs are used.

The position stability of the arc is achieved by a proper configuration of an arc chamber and electrodes and by cooling of the arc column. Cooling of the arc leads to constriction of the arc column and thus to a higher energy density and temperature in the central part of the column. In most of the generators, a gas flowing along the arc column is used for arc cooling and stabilisation. The gas is often injected tangentially into the arc chamber so that a gas swirl is formed. In addition to convective cooling, the centrifugal force helps to keep the hot arc plasma at the centre of the arc chamber. The arc discharge can also be stabilised by contact with the wall of the chamber. Heat transfer to the wall is a stabilising mechanism, but for higher currents, evaporation and ablation of the wall material may lead to wall erosion and destruction.

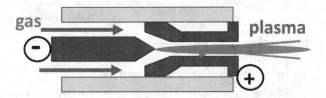

FIGURE 2.1   Schematic arrangement of plasma torch with button cathode.

Arc plasma generators can be generally divided into two classes: transferred arc torches and non-transferred plasma torches.

In non-transferred arc torches, both electrodes are parts of the torch body. In a typical arrangement of the torches for a lower power, up to 200 kW, a rod-shaped cathode is positioned at the entrance of a cylindrical nozzle anode. The plasma-forming gas is injected into the arc chamber along with the cathode. The anode serves also as an exit nozzle, which determines the shape of the generated plasma jet. A schematic arrangement of this torch type is shown in Figure 2.1.

In the alternative design, which is used mostly for higher powers and higher plasma flow rates, both the electrodes have cylindrical forms and the plasma gas is injected into the chamber through the spacing between the electrodes. Magnetic coils are often positioned around the electrodes to enhance the arc spot motion on the internal surfaces of cylindrical electrodes. The motion of arc spots leads to a reduction of local heating in the position of spots and thus to a reduction of electrodes' erosion. The schematic of this type of plasma torch is shown in Figure 2.2.

In both the described types of plasma torches, the arc discharge is stabilised by gas flowing along the discharge. The gas flowing between the arc column and the electrode walls also protects the electrode's surface against overheating.

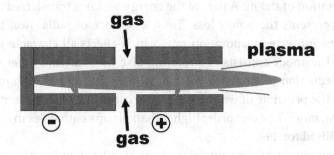

FIGURE 2.2   Schematic arrangement of plasma torch with tubular electrodes.

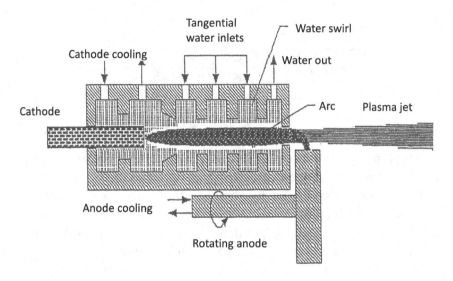

FIGURE 2.3 Schematic configuration of water-stabilised plasma torch.

In a special type of plasma torch with water-stabilised arc (Hrabovsky et al. 1997), the gas is replaced by water. A schematic picture of the water-stabilised plasma torch is shown in Figure 2.3. An arc is ignited in the centre of a vortex of water which is created in an arc chamber by means of tangential water injection. Water flows over the segments, which determine the inner diameter of the vortex, into the exhaust slots. Evaporation from the inner surface of the vortex, heating, and ionisation of steam are principal mechanisms that produce arc plasma. Energy, which is dissipated in the conducting arc core by Joule heating, is transported radially to the inner surface of the vortex by radiation, heat conduction, and turbulent transfer. Evaporation rate m is determined by a fraction of total power reaching the liquid. The other part of transferred energy is absorbed in a zone between the arc column and the surface of water leading to heating and ionisation of steam. A part of the energy which is transferred into the liquid represents the power loss. Thus, a balance of radial heat transfer is decisive for water evaporation rate, which affects all characteristics of the arc. The anode consists of a rotating disc with internal water cooling for the reduction of electrode erosion, which is extremely high in steam plasma. The principle of water-stabilised arc leads to very high enthalpies of plasma, more than one order higher than plasma enthalpies in common gas-stabilised torches.

A combination of the principles of gas-stabilised and water-stabilised arcs is used in a hybrid plasma torch (Hrabovsky et al. 2006), as shown in

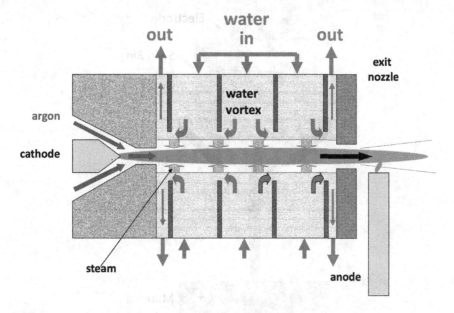

FIGURE 2.4 Schematic configuration of hybrid water/gas plasma torch.

Figure 2.4. In a cathode section with configuration of the gas-stabilised torch a gas plasma is produced, and this plasma enters the chamber with water vortex, where steam plasma is produced which is mixed with the flowing gas plasma. The principle allows the control of plasma characteristics in a broad range. This torch offers the production of plasma with properties of water-stabilised plasma and with substantial reduction of cathode erosion.

In the plasma torches described above, both electrodes, cathode and anode, are parts of the plasma torch body. In the alternative configuration, the cathode and anode parts of the torch are separated, and an anode is positioned outside the torch body. The torches are known as transferred plasma torches. The typical configuration of a plasma torch with transferred electric arc is shown in Figure 2.5. The anode is usually created by treating a material which must be electrically conductive. The main advantage of plasma torch is higher energy efficiency compared with non-transferred plasma torches and the absence of problems with anode erosion. The control of arc parameters is more difficult, especially in cases where materials are injected into the volume of reactor and the presence of material between the cathode part and the anode can substantially influence the arc behaviour.

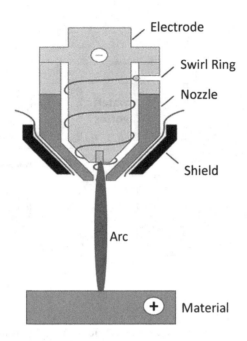

FIGURE 2.5 Plasma torch with transferred arc.

In some applications, plasma torches with AC arcs are utilised. The configuration with cylindrical electrodes is used. The main advantage is the usage of a more simple power supply without a rectifier. However, the arc is less stable, and it may be extinguished in the moments when the arc current passes zero value. The necessity of re/ignition of the arc in every zero current then makes the power supply system more complicated.

Three-phase plasma torches with arcs ignited between three electrodes are also used in some plasma waste treatment reactors.

2.2 PROCESSES AND PROPERTIES OF ARC DISCHARGE

The arc discharge ignited between two electrodes, cathode and anode, has three sections with completely different dominating processes and different types of plasma. A schematic picture of the electric arc, distribution of voltage along with the discharge, and the radial profile of plasma temperature in the arc column is shown in Figure 2.6.

Although the two electrode sections are characterised by a relatively high electric field intensity and a plasma that is not in thermal equilibrium, the plasma in the arc column is in thermal equilibrium and the electric field intensity is substantially lower than that in the electrode regions. The schematic picture of electric arc discharge, distribution of voltage

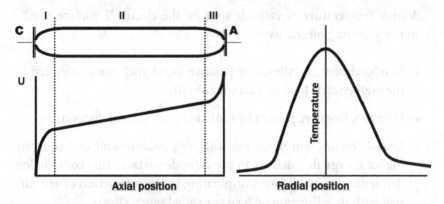

FIGURE 2.6 Schematic picture of electric arc, distribution of voltage along with the discharge, and radial profile of plasma temperature in the arc column.

along the arc length, and radial profile of plasma temperature in the arc column are shown in Figure 2.6. The dimensions of the electrode regions are in the range of the mean free path of electrons and ions and are in most cases substantially smaller than the length of the arc column, which is determined by the distance and configuration of electrodes. The arc column is the region where thermal plasma is utilised for material treatment, and the role of the electrode regions is the transfer of electric current from the discharge body to the electrodes.

2.2.1 Cathode Region

Electric current is transferred from the cathode surface into the arc body by electrons emitted from the cathode and by positive ions originating in the arc plasma. Between the cathode and the arc column is created the near-cathode sheath region which is collisionless. The electric field intensity in this sheath is enhanced due to the space charge of positive ions. The main mechanism of electron emission from the cathode is thermionic emission. Due to high electric field intensity, the emission is enhanced by a high electric field, which causes reduction and deformation of a potential barrier for electron release from the cathode (TF emission). Thermionic emission is intensive, especially in cathodes made of refractory metals (e.g., tungsten, graphite), with high melting and boiling temperatures. For cathodes made of metals with lower melting temperature (copper), the presence of high electric field intensity is important.

A high temperature of cathode spot on the electrode surface results from the following mechanisms:

- Bombardment of cathode by positive ions which are accelerated in the high electric field in a cathode sheath.
- Heat flux from arc plasma by radiation and heat conduction.
- Radial constriction of current-carrying column and consequently higher energy flux density to the cathode surface. This constriction density is caused by forces originating in the interaction of arc current with its self-magnetic field (so-called pinch effect).

Energy losses from the cathode include heat conduction into the electrode body, radiation from the cathode surface, and evaporation of the cathode material.

The temperature of the cathode surface in the cathode spot is established by the balance of energy input and energy losses. The temperature must be high enough to ensure a sufficient electron emission current.

For cathodes made of refractory metals with high melting points (tungsten, graphite), the temperature in the cathode spot is close to and sometimes higher than the melting temperature. These cathodes are called hot cathodes. Tungsten cathodes are usually of conical form, and the arc spot is at the tip of the cone. Graphite cathodes in the form of a rod are sometimes consumable, and their erosion resulting from high temperatures is compensated by moving the electrode.

Cathodes made of materials with lower melting points operate with lower cathode spot temperatures and are called cold cathodes. Thermionic emission is enhanced by the effect of a high electric field in the cathode region. The electric field can also be enhanced by the presence of small-scale unevenness on the cathode surface. The cathodes have often the form of a cylinder and the cathode spot is moved on the inner surface of the cylinder by the effect of plasma gas flow and magnetic forces acting on the cathode attachment produced by external coils positioned on the cathode cylinder.

As electrode erosion due to high temperatures in the electrode region is the principal factor determining the lifetime of the plasma torch, the control of the cathode temperature is one of the important factors of plasma torch design. The cathode body must be efficiently cooled by water cooling, on the other hand, the conditions for obtaining sufficient temperature in the cathode spot must be ensured. Intensive cooling of the cathode body and rapid movement of cathode spots in cold cathodes are solved in the electrode design. For electron emission work function of the cathode material and consequent reduction of temperature needed for sufficient electron emission, the current density is another important design factor. Tungsten doped with thorium is used as the cathode material, for lower currents up to several hundreds of amperes, the cathodes made of hafnium with low work function are fixed in the copper body.

2.2.2 Anode Region

The electric current from the arc plasma to the anode is transferred through a small-scale region with specific plasma properties substantially different from the properties of equilibrium of thermal plasma in the arc column. The depth of the region is small, typically millimetres or even less. The voltage drop in the anode region is typically several volts, depending on arc current, anode material, and plasma gas. The anode region is characterised by high-temperature gradient and high electric field. The electric current from plasma to the anode surface is primarily transferred by the electrons. Electrons are accelerated by a high electric field, and bombardment of the anode surface by electrons leads to heating of the electrode. The other mechanisms which contribute to anode heating in the location of the anode spot are radiation and heat conduction from high-temperature plasma. High temperature which may lead to the anode material melting and erosion of electrode results from high current density in the anode spot. The current-carrying column is constricted in the anode region as a result of two main factors

- Electromagnetic forces due to effect of self-magnetic field of arc current on current-carrying column.
- Presence of metallic vapour with lower ionisation potential due to evaporation of an anode material.

The size and properties of the anode region are influenced by other factors like anode material, the kind of plasma gas, and configuration of a gas flow field.

In some specific cases, the anode region could be maintained in a diffuse mode with large current-carrying zone and thus low current density. The melting and erosion of the anode surface in a diffuse mode are minimal.

In most cases, especially for a higher arc current, the diffuse mode cannot be maintained and the small-scale anode spot with high current density and anode surface heating leads to anode melting and erosion. Reduction of these effects is achieved by effective anode body cooling and by forced movement of an anode attachment region on the anode surface. In the common design of arc plasma torches, an exit nozzle serves as an anode and the movement of anode spot on the anode surface is affected in several ways:

- Tangential injection of plasma-forming gas into the torch body.
- External coils creating magnetic field acting on the anode attachment region.
- Shape of anode nozzle leading to anode attachment caused by draft forces from the gas flow and magnetic forces from self-magnetic field of arc current

The periodic movement of an anode attachment along the electrode surface is typical for DC arc plasma torches with nozzle anode. A movement of the attachment in the direction of plasma flow is followed by the formation of a new anode attachment by a restrike in a downstream position. This results in a periodic movement of the anode attachment causing changes in the arc column length with typical saw-tooth shape fluctuations of an arc voltage and power. Fluctuations and instabilities of the produced plasma jet may cause problems in some applications, but they are not important in plasma waste treatment technology. On the other hand, the movement of anode spot leads to the reduction of local overheating of anode surface and may contribute to reduction of anode erosion. As generally an anode erosion is the main factor limiting the lifetime of the whole plasma torch, all ways of reduction of anode erosion should be considered in plasma torch design.

2.2.3 Arc Column

The arc column is an arc region, where thermal plasma which is utilised for material treatment is produced. In arc torches, thermal plasma produced in an arc column flows through the exit torch nozzle out of the torch body.

Plasma in the arc column of high current arcs is in the local thermodynamic equilibrium (LTE). Temperatures of all components of plasma, i.e., electrons, ions, and neutral particles, are equal or the temperature differences are negligible. Equilibrium in plasma is achieved through an intensive interaction of all plasma components. Elastic as well as inelastic collisions between plasma particles are the main mechanisms determining the plasma properties. This is opposite to the processes in cathode and anode regions which are almost collisionless. Thermodynamic equilibrium in a plasma of a column of high current arc results from high particle densities, high temperature, and relatively low electric field intensity in the arc column. Electrons which are accelerated in an electric field suffer from a high number of collisions with other particles in the plasma (molecules, atoms, and ions), and the kinetic energy of accelerated electrons is transferred to all particles in these collisions. Thus, LTE in plasma is achieved under conditions with lower electric field, high pressure, and high temperature in the plasma.

Plasma in LTE can be described as a fluid with defined thermodynamic and transport properties which are dependent on the composition of plasma-forming gas, temperature, and pressure. The composition and material properties of plasma for a given plasma-forming gas can be calculated using the laws of thermodynamics.

Figure 2.7 presents the calculated composition of argon plasma in LTE at atmospheric pressure as a function of plasma temperature.

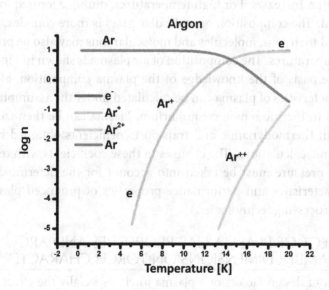

FIGURE 2.7 Composition of argon plasma at atmospheric pressure.

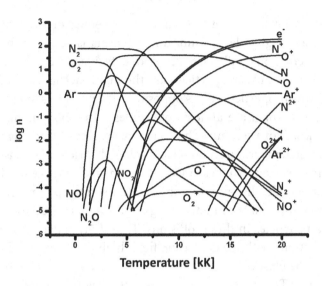

FIGURE 2.8   Composition of air plasma at atmospheric pressure.

For monoatomic gases like argon, a plasma is composed of gas atoms and ions and electrons created by the ionisation of atoms. With an increase in temperature, the kinetic energy of chaotic thermal movement of particles increases, and for some particles it is higher than the energy needed for the ionisation of an atom. The number of ionisation collisions increases with an increase in temperature and consequently the degree of ionisation increases. For high temperatures, multiple ionised ions may be formed. The composition of molecular gases is more complex, besides atoms and their ions, molecules and molecular ions may also be present at lower temperatures. The composition of air plasma is shown in Figure 2.8.

On the basis of the knowledge of the plasma composition, all material characteristics of plasma can be calculated under the assumption that plasma is in thermodynamic equilibrium. Plasma can be then treated as a gas with thermodynamic and transport coefficients obtained by thermodynamic calculations. The changes in these coefficients with temperature and pressure must be taken into account for the determination of arc characteristics and performance properties of produced plasmas in plasma processing technologies.

2.3 EFFECT OF PLASMA GAS PROPERTIES AND ARC CHAMBER DIMENSIONS ON TORCH CHARACTERISTICS

The effect of design factors of a plasma torch, especially the effect of configuration and dimensions of arc chamber and the kind of flow rate of plasma-forming gas, can be analysed based on arc models. Numerical models are based on a solution of equations of balances of energy, momentum, and mass in plasma flows. Plasma is described in these models as a fluid with thermodynamic and transport properties calculated for a given plasma gas.

In the following, a simplified analytic model is described which gives basic information about the effect of design factors on arc properties. The model describes an arc in a cylindrical arc chamber which is stabilised by the axial flow of plasma gas. The model includes also the effect of radial inflow of stabilising gas from the walls of the cylindrical chamber. Thus, it may be also used for arc description in water-stabilised plasma torches.

In the arc channel model, an arc column is considered as the cylindrical plasma column with the same temperature in the whole volume of the column. The column is in the cylindrical arc chamber with fixed wall temperature. The energy balance equation of this system is obtained by an integration of the exact three-dimensional energy equation. The following energy balance is considered (Figure 2.9):

The integral energy balance equation of a cylindrical arc column can be written in the form [9]:

$$\frac{\partial(\overline{\rho v_z h}A)}{\partial z} = A\overline{\sigma}E^2 + 2\pi R\left(k\frac{\partial T}{\partial r}\right)_{r=R} - 4\pi\overline{\varepsilon_n}A, \qquad (2.1)$$

where ρ is the plasma density, v_z the axial velocity, h the enthalpy, σ the electric conductivity, k the thermal conductivity, T the temperature, ε_n the net emission coefficient representing power loss due to radiation, and E the electric field intensity. $A = \pi R^2$ is the cross section of the arc chamber. Equation (2.1) was obtained from the energy balance equation of the cylindrical arc column which is integrated over the cross section of arc chamber. Quantities averaged over the cross section A are defined by the equation (for a quantity X):

$$\overline{X} = \frac{1}{\pi R^2}\int_0^R 2\pi r X dr. \qquad (2.2)$$

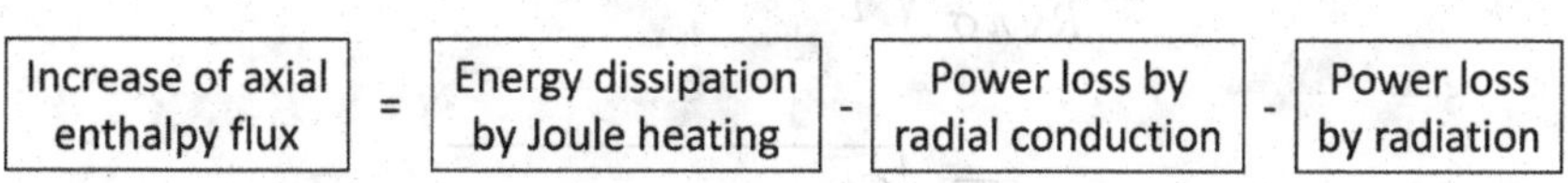

| Increase of axial enthalpy flux | = | Energy dissipation by Joule heating | − | Power loss by radial conduction | − | Power loss by radiation |

FIGURE 2.9 Energy balance in the arc column.

The simple equation can be obtained if derivatives in Equation (2.1) are approximated as

$$\frac{\partial (\overline{\rho v_z h} A)}{\partial z} = \frac{\overline{\rho v_z h} A}{L}, \qquad (2.3)$$

and

$$\left(k\frac{\partial T}{\partial r}\right)_{r=R} = \frac{\partial \overline{S}}{\partial r} = -\frac{\overline{S}}{R}, \qquad (2.4)$$

where L is the length of the arc chamber, R its radius, and heat flux potential S is defined as

$$S = \int_{T_0}^{T} k\, dT. \qquad (2.5)$$

The integral energy balance equation can then be written in the form:

$$\frac{G\overline{h}}{L} + 2\pi \overline{S} + 4\pi^2 R^2 \overline{\varepsilon_n} = \pi R^2 \overline{\sigma E^2} = \frac{I^2}{\pi R^2 \overline{\sigma}}, \qquad (2.6)$$

where the arc current and the total mass flow rate are given by the equations

$$I = E\int_0^R 2\pi r\sigma\, dr = \pi R^2 \overline{\sigma} E, \qquad (2.7)$$

$$G = \int_0^R 2\pi r\rho v_z dr = \pi R^2 \overline{\rho v_z}. \qquad (2.8)$$

The following two equations describing the relations between electric arc parameters I, E, and the material properties and flow rate of plasma gas can then be derived

$$E = \frac{1}{R\sqrt{\pi \overline{\sigma}}}\sqrt{\frac{G\overline{h}}{L} + 2\pi \overline{S} + 4\pi^2 R^2 \overline{\varepsilon_n}}, \qquad (2.9)$$

$$I = R\sqrt{\pi \overline{\sigma}}\sqrt{\frac{G\overline{h}}{L} + 2\pi \overline{S} + 4\pi^2 R^2 \overline{\varepsilon_n}}. \qquad (2.10)$$

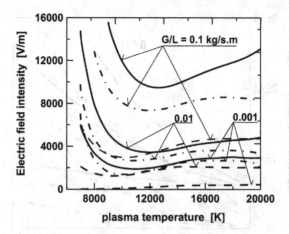

FIGURE 2.10 Electric field intensity in the arc column with radius $R = 3$ mm for different ratios of gas flow rate G to the arc length L for arc in steam (full line), nitrogen (dash-and-dot line), and argon (dashed line).

On the right-hand side of Equations (2.9) and (2.10) are the material coefficients of plasma σ, h, S, and ε_n which are dependent on the plasma temperature, R and L are the geometrical design parameters, and G is the total mass flow rate of plasma gas. Thus, relations between electric characteristics of the arc I, E, and averaged plasma temperature can be found in Equations (2.9) and (2.10).

Figure 2.10 presents the relation between the plasma temperature and electric field intensity in an arc column for three plasma gases. It can be seen how the choice of plasma gas influences the electric field intensity and thus arc power. High electric field intensity and thus high power can be obtained in plasma torches operated with steam, while plasma torches with argon as plasma gas have substantially lower power for the same arc current.

The dependence of plasma enthalpy on arc current is shown in Figure 2.11. The highest plasma enthalpies for a given current are in an arc with steam. Thus, steam torches are characterised by high power and high capacity to accumulate energy and transport it to treated material in plasma processing technologies. Limitations in the application of steam in plasma torches are given by the high erosion of electrodes in steam plasma arcs.

Torch power can be calculated as $I.E.L$, and the torch efficiency as the ratio of energy carried by plasma flow through the exit torch nozzle to the energy transported radially from the arc column to the arc chamber wall.

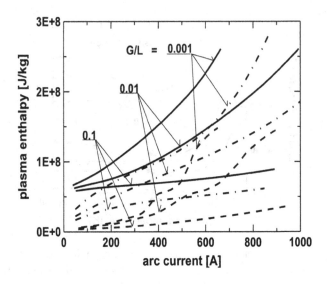

FIGURE 2.11 Dependence of mean plasma enthalpy on the arc current for different ratios G/L for arc in steam (full line), nitrogen (dash-and-dot line), and argon (dashed line).

The following equations can be derived for the efficiency η and the torch power P

$$\eta = \left(1 + 4\pi^2 \frac{R^2 L}{G} \frac{\varepsilon_n}{h} + 2\pi \frac{L}{G} \frac{S}{h}\right)^{-1}, \quad (2.11)$$

$$P = I \cdot U = I \cdot \left(\frac{1}{\eta} \frac{L \cdot G}{\pi R^2} \frac{h}{\sigma}\right)^{\frac{1}{2}}. \quad (2.12)$$

For a given arc current, the torch power as well as the torch efficiency can be increased by an increase of gas flow rate G. The decisive plasma gas properties for the torch power and the torch efficiency are the enthalpy h, the heat flux potential S (Equation 2.5) depending on thermal conductivity, and the net emission coefficient ε_n characterising power loss by radiation. The heat flux potential S and the net emission coefficient ε_n are the integral quantities that determine energy transfer from arc column to the wall, and plasma enthalpy is the characteristic capacity of plasma to accumulate energy. High plasma gas enthalpy results in high torch power.

The high enthalpy of steam plasma (Figure 2.11), together with high thermal conductivity, and steam plasma composition (atoms and ions of hydrogen and oxygen) result in advantages of applications of steam plasma torches in plasma gasification technology. The limitations of the usage of steam as plasma gas are due to the high erosion rate of arc electrodes in steam plasma and from technical problems of steam generation. The erosion of electrodes is low in argon plasma, but the low value of plasma enthalpy leads to low arc power. Most plasma systems, especially for high power, are operated with air.

REFERENCES

Boulos, Maher, Pierre Fauchais, and Emil Pfender. 1994. *Thermal Plasma Fundamentals and Applications.* New York: Plenum Press.

Hrabovský, Milan, Miloš Konrád, V. Kopecky, and Viktor Sember. 1997. "Processes and Properties of Electric Arc Stabilized by Water Vortex." *IEEE Transactions on Plasma Science* 25, no. 5: 833–839.

Hrabovsky, Milan, V. Kopecky, Viktor Sember, Tetyana Kavka, Oleksiy Chumak, and Milos Konrad. 2006. "Properties of Hybrid Water/Gas DC Arc Plasma Torch" *IEEE Transactions on Plasma Science* 34, no. 4, Part 3: 1566–1575.

CHAPTER 3

Conventional Thermochemical Technologies for Waste Treatment

Michal Jeremiáš

THERMAL TREATMENT OF WASTE is a proven way to minimise it. Traditionally, wastes have been thermally reduced to ash without the use of their bound energy in order to reduce the volume of material to be landfilled. At present, however, most of the thermally treated waste is used for electricity generation or for combined heat and power generation in incinerators or waste-to-energy plants. Combustion technologies are very advanced, reliable, and significantly predominant in the thermal processes of waste processing, especially grate incineration (Lombardi, Carnevale, and Corti 2015; Leckner 2015). However, as will be shown in the following text, conventional incinerators have limits in achievable efficiency and scaling down so that the amount of treated waste corresponds to the real agglomeration of a smaller city. These limits can be circumvented by using innovative technologies (Lamers et al. 2013), which may include technologies using gasification and pyrolysis.

Using gasification, the thermochemical conversion of fuel to heating gas takes place. This conversion is made possible by the action of gasification

DOI: 10.1201/9781003096887-3

medium and high temperature. Air is usually used as the gasification medium, but in special cases steam or a mixture of oxygen and steam or carbon dioxide can also be used. The product is a gas containing heating components (H_2, CO, CH_4, and other organic minor compounds), accompanying components (N_2, CO_2, and H_2O), and pollutants (tar, dust, sulphur compounds, chlorine compounds, alkali, and others). Depending on the degree of gasification, the solid residue is ash or a carbonaceous porous residue. The produced heating gas can be used for various applications, the relatively easiest one being to combust it in a boiler to produce energy steam. More complicated uses include the propulsion of a gas engine or turbine for electricity production or even the synthesis of chemicals (especially alternative fuels), and in all these applications it is essential to properly clean the gas, especially (but not only) from tar compounds, which makes this technology complicated (Pohořelý et al. 2012; Pohořelý et al. 2014; Jeremiáš et al. 2014). There are many concepts and demonstration units for waste gasification and pyrolysis, and they are described in detail in review articles (Malkow 2004; Arena 2012; Arena 2013). Even though these are interesting and technologically feasible processes, it should be noted that they are not sufficiently mature in comparison with conventional incinerators. Nevertheless, there are examples of successful, economically viable, and proven technologies for electricity and heat production from waste that use gasification or pyrolysis technology. Two successful examples will be presented in this chapter.

3.1 WASTE INCINERATION

For the purposes of this text, it is necessary to summarise a few facts about waste incinerators. Incinerators are relatively small compared to conventional power plants and heating plants (37.5% of incinerators process <100,000 t/year of waste, 39.5% of 100–250,000 t/year, and 39.5% of >250,000 t of waste per year). The average European values of electrical efficiency in 2009 were 20.7% for equipment operating in condensing mode and 14.2% for equipment operating in combined heat and power mode (Reimann 2009; Reimann 2013). These values of electricity generation efficiency are relatively small compared to conventional coal-fired power plants. The main reason, in addition to a smaller scale, is the need to prevent significant high-temperature acid corrosion caused by metal chlorides in the fly ash and high concentrations of hydrogen chloride (HCl) in the flue gas (Lee, Themelis, and Castaldi 2007; Persson et al. 2007; De Greef et al. 2013). The degree of high-temperature corrosion depends mainly on the surface temperature of the metal; thus, to prevent high-temperature corrosion, the surface temperature of evaporators and superheaters must be limited by limiting the evaporating pressure and superheating temperature. In addition, it is necessary to prevent the risk of condensation during steam expansion in steam turbines, which in turn means an additional limitation for the selected evaporating pressure (Lombardi, Carnevale, and Corti 2015). For these reasons, conservative steam parameters are usually chosen (usually the maximum superheating temperature is around 400°C) together with the limitation of the flue gas temperature in the superheater below 650°C (De Greef et al. 2013), which also results in a lower efficiency of electricity generation. In modern incinerators, higher steam parameters are sometimes achieved (60 bar and 500°C or even higher), but this leads to the need to use corrosion-resistant materials (e.g., Inconel 625) to protect the exchangers, which also results in increased investment costs, but this pays off for larger devices (Lombardi, Carnevale, and Corti 2015). The possibility to increase the electrical efficiency is the reheating of steam, which is again financially feasible only for large devices (Pavlas et al. 2011).

The second cause of the reduction in electrical efficiency is the high chimney loss. This tends to be in the range of 7–25% of the energy of the fuel (waste), depending on the flue gas outlet temperature and the excess air during combustion (Pavlas et al. 2011). The flue gas temperatures at the outlet are usually cooled by an exchanger system to 150–380°C (Stehlík 2012). The excess air is usually in the range of 1.75–1.9, but by optimising the combustion chamber it is possible to reduce it to 1.39 (Gohlke 2009; Lombardi, Carnevale, and Corti 2015). Another way to reduce these losses (to a limited extent) is to use flue gas recirculation to control the temperature of the combustion chamber at the inlet of fresh air to achieve an oxygen concentration in the flue gas as close to 6% as possible (Lombardi, Carnevale, and Corti 2015).

An example of a modern large incinerator with high electrical efficiency is the incinerator in Amsterdam, which was put into operation in 2007. This plant with a horizontal grate burns 530,000 t of waste per year with an average calorific value of 10 MJ kg^{-1}. A high electrical efficiency of >30% is achieved by high steam parameters (440°C and 130 bar) and its reheating behind the first stage of the turbine at a pressure of 14 bar to 320°C saturated steam. The pressure in the condenser is maintained at 0.03 bar by cooling with seawater. The availability of the facility in 2011 was approximately 93%. The excess air in this incinerator is 1.4 and approximately 25% of the flue gas is recirculated (Murer et al. 2011).

3.2 LARGE-SCALE WASTE GASIFICATION

As can be seen from the above, the main reasons for a lower efficiency of electricity generation in incinerators are (1) high-temperature corrosion caused mainly by chlorine compounds in the flue gas and the related need to keep steam parameters rather conservative and (2) high chimney loss caused mainly by high excess air required for sufficient waste incineration. These two causes can be circumvented by dividing the waste incineration process into (i) a gasification step, (ii) a purification of the resulting gas above the dew point of tar and below the dew point of alkaline corrosive chlorine compounds, and (iii) its combustion with a minimal excess of air.

This concept is successfully demonstrated at the Kymijärvi II facility in Lahti (Finland). The city of Lahti also became the winner of the 2021 European Green Capital competition due to the running of this facility (Fleming 2021). The Kymijärvi II facility has been in operation since the end of 2011, and it processes 250,000 t of refuse-derived fuel (RDF) fuel per year and produces 50 MW of electricity and 90 MW of heat in cogeneration mode from 160 MW of fuel input. This corresponds to an electrical efficiency of 31% and an overall efficiency of 87.5% (Bolhar-Nordenkampf and Isaksson 2016). RDF fuel in this case means a mixture of waste from industry, retail, construction, and household, which is not suitable for recycling. This waste is collected from a radius of 200 km, which includes the city of Helsinki, which is 100 km away. The purchase of fuel is ensured by long-term contracts and its quality is carefully verified (Ricardo-AEA 2013). The fuel is further pretreated by grinding and removing metals in magnetic separators and eddy current separators. The calorific value of the fuel is in the range of 8–15 MJ kg^{-1} (Bolhar-Nordenkampf and Isaksson 2016). The plant has a successful operation of more than 26,000 h with 20,000 h in operation with RDF and 6,000 hours with a mixture of RDF and waste timber. The construction of the facility began in the spring of 2010, and at the end of 2011 the facility was already in trial operation. In the spring of 2012, the equipment was handed over for commercial operation. The total cost of this facility was approximately €160 million. The average availability during the first year of operation was 80%, and after the subsequent improvement of the function of high-temperature ceramic filters, the availability increased to 90%. The fuel is gasified in two fluidised-bed generators, each with an input of 80 MW in the fuel, and the raw gas produced (850–900°C) is cooled to 400°C using heat to preheat the boiler water (Bolhar-Nordenkampf and Isaksson 2016). The fluidised bed of the generators consists of a mixture of sand and limestone

(Ricardo-AEA 2013). When the gas is cooled to 400°C, the alkaline corrosive chlorine compounds condense on solid particles in the gas, which are filtered off on ceramic filters at this temperature. At a temperature of 400°C, tar remains present in the gaseous (non-condensed) form. The gas thus purified (containing tar vapours) is subsequently burnt in a boiler with a very low excess of air to produce steam with parameters of 121 bar and 540°C. Such high steam parameters (compared to 40 bar and 400°C in a conventional incinerator) make it possible to achieve the above-mentioned high electrical efficiency of 31% with respect to the fuel input (Bolhar-Nordenkampf and Isaksson 2016). The flue gases from the boiler undergo multi-stage treatment (denitrification, dedusting, sodium bicarbonate injection, and also limestone with activated carbon), thanks to which emissions of pollutants in the flue gases are minimal (CPGA 2022a).

Of course, it must be added that this plant in Lahti is a clear exception, and several similar attempts to gasify waste in the past have failed. For example, a similar plant in Greve in Chianti, Italy, for processing 200 t of RDF per day in the form of pellets was shut down after three periods of operation over five years (Morris and Waldheim 1998; Belgiorno et al. 2003; Lombardi, Carnevale, and Corti 2015).

3.3 USE OF GASIFICATION TECHNOLOGY FOR WASTE TREATMENT ON A SMALLER SCALE

Another reason for the use of gasification in waste treatment may be the effort to ensure minimum emissions of the plant even on a smaller scale suitable for the capacity of a smaller urban agglomeration. This avoids the need to transport waste from long distances to a large facility and thus avoids the emissions and costs associated with waste transport. Also, the amount of heat produced by the thermal utilisation of local waste is easier to apply within the local heat distribution network.

Energos technology is based on this logic ('Energos' 2022; CPGA 2022b). This technology is said to achieve low emissions of carbon monoxide (CO), nitrogen oxides (NOx), and dioxins without the use of standard gas cleaning technologies used in traditional incinerators. Also, the content of carbon (non-burnt) in the ash is below 3% total organic carbon (TOC) and its leachability is at the level of 10% compared to conventional equipment for energy recovery of waste ('Energos' 2022; Pugh, Read, and Mitchell 2011).

Energos technology was developed at SINTEF in Trondheim, Norway, between 1990 and 1997, and over the next five years six heating plants were built. They are still in operation (five in Norway and one in Germany), each processing 39,000 t/year of residual non-recyclable municipal and industrial waste. In 2004, the energy company Energ-G acquired the technology of the Norwegian company Energos ASA and brought the technology to the UK market, where a unit was built on the Isle of Wight in 2009 (30,000 t/year) (Pugh, Read, and Mitchell 2011). Since then, another unit has been put into operation in Norway in 2010 (78,000 t/year) and three more are currently under construction in the United Kingdom (144,000, 96,000, and 144,000 t/year) ('Energos' 2022).

The technology is intended for energy recovery of residual non-recyclable municipal and industrial waste and its main advantage is very low emissions of nitrogen oxides and dioxins, without the need for secondary measures (in the case of NOx emissions, to reach values at the level of about 20% of legal limits, for dioxins it is even 1%). The above is made possible by the application of the principle of gasification on the grate. Before gasification, the fuel is ground, and metals are separated on a magnetic separator. The grate is designed as horizontal and movable, and gasification takes place in the absence of oxygen (average stoichiometric coefficient of excess air in this chamber is about 0.5 and air is distributed to the grate to achieve a minimum concentration of carbon in the ash) at temperatures around 900°C. The heating gas (containing 14% CO, 5% H_2, and 4% CH_4) then passes through an oxidation chamber (with the addition of secondary air and recirculated flue gases), it is dedusted and is then used to produce heat in a steam boiler (23 bar, 380°C) and electricity depending on the heat demand, which is the primary product in the case of this technology. Thanks to its construction, the flue gases in the boiler are rapidly cooled below temperatures that preclude the recombination of dioxins and then, with the addition of ground limestone and activated carbon, freed of other monitored pollutants in the fabric filter (CPGA 2022b; Pugh, Read, and Mitchell 2011; del Alamo et al. 2012; Ellyin and Themelis 2011; 'Energos' 2022; Ellyin 2012).

3.4 FURTHER POSSIBLE DEVELOPMENT OF TECHNOLOGIES USING WASTE GASIFICATION AND PYROLYSIS

Waste gasification and pyrolysis technologies have a future in applications with the greatest possible simplicity. This can be achieved in fluidised-bed reactors via autothermal operation with air being the gasifying agent while producing a low-calorific gas (typically 3–5 MJ/m^3). This configuration allows the processing of fuel with a slightly variable composition, and it is possible to use a cheap natural catalyst such as limestone or dolomite (in calcined forms) as the fluidised-bed material (Pohořelý et al. 2016; Corella, Toledo, and Molina 2008). The use of fluidised-bed gasification of waste with subsequent purification of gas from corrosive compounds and heavy metals with subsequent combustion of this gas in existing fossil fuel boilers (with the possibility of producing high-parameter steam) or in cement kilns and kilns for lime production. In this way, part of the fossil fuel in the boiler could theoretically be replaced by low-calorific gas made from waste. Co-combustion of the gas is likely to have minimal impact on the operation of existing installations, corrosion, ash, and emissions. On the contrary, from experience with a similar concept from the Vaasa plant (Finland), it can be concluded that it is possible to improve the emission characteristics of the boiler. The investments of the outlined concept would be significantly lower than in the construction of a whole new facility for energy recovery of waste (Bolhar-Nordenkampf and Isaksson 2016). While maintaining this simple concept, it is promising to explore the possibilities of high-temperature gas cleaning (Svoboda et al. 2017). Also, finding the ideal pretreatment of real waste for gasification and testing the possibility of synergistic co-gasification of another fuel (e.g., waste wood biomass) together with waste is an interesting option (Ramos et al. 2018; Balas et al. 2017). Verification of the possibilities of using the generated gas for more demanding applications (e.g., gas engine with a generator for electricity production) should be conducted. The use of gasification media other than air, specifically H_2O and CO_2 together with oxygen (with a stoichiometric air coefficient of around 0.3), interestingly increased the usability of the produced gas (Jeremiáš et al. 2014; Pohořelý et al. 2014) for the production of second-generation biofuels or chemicals (Pohořelý et al. 2012; Fatih Demirbas, Balat, and Balat 2011; Sikarwar et al. 2017).

In recent years, interest in the application of advanced thermal plasma technology for the treatment of wastes has increased enormously, primarily because of its ability to supply elevated temperatures with high energy densities, independence from the type of waste, potential to generate saleable co-products coupled with a high destruction efficiency, control of the processing environment, environmental compatibility, and energy recovery (Sikarwar et al. 2020). In favour of thermal plasma gasification (or plasma pyrolysis) plays the minimal concentration of tar compounds in the produced calorific gas (or syngas). This is true if the process is carried out in a way that the produced gas leaves the reaction zone at high temperatures (>1000°C). In comparison to conventional gasification pathways (such as fluidised-bed gasification), the produced synthesis gas does contain only traces of tars (tens of mg/m^3) (Agon et al. 2016) as compared to tens of g/m^3 in fluidised-bed gasification (Jeremiáš et al. 2018), which renders possible the cleaning of syngas at temperatures 100–200°C without the risk of tar condensation on the filters. This temperature range can be advantageously used for gas cleaning via adsorption of catalytic poisons (such as HCl) directly on the filter dusted with suitable adsorbents (e.g., sodium bicarbonate), thus enhancing the use of the gas in demanding applications such as methanol synthesis. The flip side of plasma gasification is the consumption of electricity for providing the energy for the allothermal process. But there is a strong possibility to adjust the power of the process to actual electricity prices in the distribution grid on account of strong flexibility of the process, and therefore, it can be used as a means of transformation of surplus electricity into energy of hydrogen or chemicals, such as methanol or alternative transport fuels (Sikarwar et al. 2021).

REFERENCES

Agon, N., Milan Hrabovský, O. Chumak, Michal Hlína, Vladimír Kopecký, Alan Mašláni, A. Bosmans, et al. 2016. "Plasma Gasification of Refuse Derived Fuel in a Single-Stage System Using Different Gasifying Agents." *Waste Management* 47: 246–255. doi:10.1016/j.wasman.2015.07.014.

Arena, Umberto. 2012. "Process and Technological Aspects of Municipal Solid Waste Gasification. A Review." *Waste Management* 32 (4): 625–639. doi:10.1016/j.wasman.2011.09.025.

Arena, Umberto. 2013. *Fluidized Bed Gasification. Fluidized Bed Technologies for Near-Zero Emission Combustion and Gasification*. Woodhead Publishing Limited. https://www.elsevier.com/books/fluidized-bed-technologies-for-near-zero-emission-combustion-and-gasification/scala/978-0-85709-541-1.

Balas, Marek, Martin Lisy, Petr Kracik, and Jiri Pospisil. 2017. "Municipal Solid Waste Gasification Within Waste-to-Energy Processing." *MM Science Journal* 2017 (02): 1783–1788. doi:10.17973/MMSJ.2017_03_2016137.

Belgiorno, V., G. De Feo, C. Della Rocca, and R. M. A. Napoli. 2003. "Energy from Gasification of Solid Wastes." *Waste Management* 23 (1): 1–15. doi:10.1016/S0956-053X(02)00149-6.

Bolhar-Nordenkampf, Markus, and J. Isaksson. 2016. "Operating Experiences of Large Scale CFB-Gasification Plants for the Substitution of Fossil Fuels." In *European Biomass Conference and Exhibition Proceedings*, Monday, 6 June, 2016 to Thursday, 9 June, 2016, Amsterdam, The Netherlands, 375–381.

Corella, José, José M. Toledo, and Gregorio Molina. 2008. "Performance of CaO and MgO for the Hot Gas Clean Up in Gasification of a Chlorine-Containing (RDF) Feedstock." *Bioresource Technology* 99 (16): 7539–7544. doi:10.1016/j.biortech.2008.02.018.

CPGA. 2022a. "Technologie Zplyňování METSO." http://cpga.cz/aplikovane-technologie/metso.
CPGA. 2022b. "Technologie Zplyňování ENERGOS (pro Energetické Využití Směsných Komunálních Odpadů)." http://cpga.cz/aplikovane-technologie/energos.
De Greef, J., K. Villani, J. Goethals, H. Van Belle, J. Van Caneghem, and C. Vandecasteele. 2013. "Optimising Energy Recovery and Use of Chemicals, Resources and Materials in Modern Waste-to-Energy Plants." *Waste Management* 33 (11): 2416–2424. doi:10.1016/j.wasman.2013.05.026.
del Alamo, G., A. Hart, A. Grimshaw, and P. Lundstrøm. 2012. "Characterization of Syngas Produced from MSW Gasification at Commercial-Scale ENERGOS Plants." *Waste Management* 32 (10): 1835–1842. doi:10.1016/j.wasman.2012.04.021.
Ellyin, Claudine. 2012. *Small Scale Waste-to-Energy Technologies*. Columbia University, New York.
Ellyin, Claudine, and Nickolas J. Themelis. 2011. "Small Scale Waste-to-Energy Technologies." In *19th Annual North American Waste-to-Energy Conference, NAWTEC19*, 169–176. Lancaster, PA.
"Energos." 2022. http://www.energos.com.
Fatih Demirbas, M., Mustafa Balat, and Havva Balat. 2011. "Biowastes-to-Biofuels." *Energy Conversion and Management* 52 (4): 1815–1828. doi:10.1016/j.enconman.2010.10.041.
Fleming, Sean. 2021. "This Finnish Town Will Be Carbon Neutral by 2025. Here's How." https://www.weforum.org/agenda/2021/05/finland-carbon-neutral-2025/.
Gohlke, Oliver. 2009. "Efficiency of Energy Recovery from Municipal Solid Waste and the Resultant Effect on the Greenhouse Gas Balance." *Waste Management & Research* 27 (9): 894–906. doi:10.1177/0734242X09349857.
Jeremiáš, Michal, Michael Pohořelý, Petra Bode, Siarhei Skoblia, Zdeněk Beňo, and Karel Svoboda. 2014. "Ammonia Yield from Gasification of Biomass and Coal in Fluidized Bed Reactor." *Fuel* 117 (October): 917–925. doi:10.1016/j.fuel.2013.10.009.
Jeremiáš, Michal, Michael Pohořelý, Karel Svoboda, Siarhei Skoblia, Zdeněk Beňo, and Michal Šyc. 2018. "CO_2 Gasification of Biomass: The Effect of Lime Concentration in a Fluidised Bed." *Applied Energy* 217 (November 2017): 361–368. doi:10.1016/j.apenergy.2018.02.151.
Lamers, Frans, Edmund Fleck, Luciano Pelloni, and Bettina Kamuk. 2013. "Alternative Waste Conversion Technologies." ISWA - The International Solid Waste Association, Rotterdam, 35.
Leckner, Bo. 2015. "Process Aspects in Combustion and Gasification Waste-to-Energy (WtE) Units." *Waste Management* 37: 13–25. doi:10.1016/j.wasman.2014.04.019.
Lee, Shang Hsiu, Nickolas J. Themelis, and Marco J. Castaldi. 2007. "High-Temperature Corrosion in Waste-to-Energy Boilers." *Journal of Thermal Spray Technology* 16 (1): 104–110. doi:10.1007/s11666-006-9005-4.

Lombardi, Lidia, Ennio Carnevale, and Andrea Corti. 2015. "A Review of Technologies and Performances of Thermal Treatment Systems for Energy Recovery from Waste." *Waste Management* 37: 26–44. doi:10.1016/j.wasman.2014.11.010.

Malkow, Thomas. 2004. "Novel and Innovative Pyrolysis and Gasification Technologies for Energy Efficient and Environmentally Sound MSW Disposal." *Waste Management* 24 (1): 53–79. doi:10.1016/S0956-053X(03)00038-2.

Morris, M., and L. Waldheim. 1998. "Energy Recovery from Solid Waste Fuels Using Advanced Gasification Technology." *Waste Management* 18 (6–8): 557–564. doi:10.1016/S0956-053X(98)00146-9.

Murer, Martin J., Hartmut Spliethoff, Chantal M. W. de Waal, Saskia Wilpshaar, Bart Berkhout, Marcel A. J. van Berlo, Oliver Gohlke, and Johannes J. E. Martin. 2011. "High Efficient Waste-to-Energy in Amsterdam: Getting Ready for the Next Steps." *Waste Management & Research* 29 (10_suppl): S20–S29. doi:10.1177/0734242X11413334.

Pavlas, Martin, Michal Touš, Petr Klimek, and Ladislav Bébar. 2011. "Waste Incineration with Production of Clean and Reliable Energy." *Clean Technologies and Environmental Policy* 13 (4): 595–605. doi:10.1007/s10098-011-0353-5.

Persson, Kristoffer, Markus Broström, Jörgen Carlsson, Anders Nordin, and Rainer Backman. 2007. "High Temperature Corrosion in a 65 MW Waste to Energy Plant." *Fuel Processing Technology* 88 (11–12): 1178–1182. doi:10.1016/j.fuproc.2007.06.031.

Pohořelý, Michael, Michal Jeremiáš, Petra Kameníková, Siarhei Skoblia, Karel Svoboda, and Miroslav Punčochář. 2012. "Biomass Gasification." *Chemické Listy* 106: 264–274.

Pohořelý, Michael, Michal Jeremiáš, Siarhei Skoblia, Zdeněk Beňo, Michal Šyc, and Karel Svoboda. 2016. "Transient Catalytic Activity of Calcined Dolomitic Limestone in a Fluidized Bed during Gasification of Woody Biomass." *Energy & Fuels* 30 (5): 4065–4071. doi:10.1021/acs.energyfuels.6b00169.

Pohořelý, Michael, Michal Jeremiáš, Karel Svoboda, Petra Kameníková, Siarhei Skoblia, and Zdeněk Beňo. 2014. "CO_2 as Moderator for Biomass Gasification." *Fuel* 117 (January): 198–205. doi:10.1016/j.fuel.2013.09.068.

Pugh, Michael, Adam Read, and Desmond Mitchell. 2011. "The Energos Gasification Plant: Early Performance Assessment." *Proceedings of the Institution of Civil Engineers - Waste and Resource Management* 164 (3): 191–203. doi:10.1680/warm.2011.164.3.191.

Ramos, Ana, Eliseu Monteiro, Valter Silva, and Abel Rouboa. 2018. "Co-Gasification and Recent Developments on Waste-to-Energy Conversion: A Review." *Renewable and Sustainable Energy Reviews* 81 (March 2017): 380–398. doi:10.1016/j.rser.2017.07.025.

Reimann, Dieter O. 2009. *CEWEP Energy Report II (Status 2004–2007)*. https://www.cewep.eu/wp-content/uploads/2013/01/402_Hand_out_Energy_Efficiency_Report_March_2009_final.pdf.

Reimann, Dieter O. 2013. *CEWEP Energy Report III (Status 2007–2010)*.

RICARDO-AEA. 2013. Lahti Gasification Facility, Finland. https://www.green-industries.sa.gov.au/__media_downloads/165467/Case%20Study%201%20Lahti%20Gasifier%20Finland%20FINAL.pdf.

Sikarwar, Vineet Singh, Milan Hrabovský, Guido van Oost, Michael Pohořelý, and Michal Jeremiáš. 2020. "Progress in Waste Utilization via Thermal Plasma." *Progress in Energy and Combustion Science* 81 (November): 100873. doi:10.1016/j.pecs.2020.100873.

Sikarwar, Vineet Singh, Annika Reichert, Michael Pohorely, Erik Meers, Newton Libanio Ferreira, and Michal Jeremiáš. 2021. "Equilibrium Modeling of Thermal Plasma Assisted Co-Valorization of Difficult Waste Streams for Syngas Production." *Sustainable Energy & Fuels* 5 (18): 4650–4660. doi:10.1039/D1SE00998B.

Sikarwar, Vineet Singh, Ming Zhao, Paul S. Fennell, Nilay Shah, and Edward John Anthony. 2017. "Progress in Biofuel Production from Gasification." *Progress in Energy and Combustion Science* 61 (July): 189–248. doi:10.1016/j.pecs.2017.04.001.

Stehlík, Petr. 2012. "Up-to-Date Technologies in Waste to Energy Field." *Reviews in Chemical Engineering* 28 (4–6): 223–242. doi:10.1515/revce-2012-0013.

Svoboda, Karel, Jindřich Leitner, Jaromír Havlica, Miloslav Hartman, Michael Pohořelý, Jiří Brynda, Michal Šyc, Yau-Pin Chyou, and Po-Chuang Chen. 2017. "Thermodynamic Aspects of Gasification Derived Syngas Desulfurization, Removal of Hydrogen Halides and Regeneration of Spent Sorbents Based on $La_2O_3/La_2O_2CO_3$ and Cerium Oxides." *Fuel* 197 (June): 277–289. doi:10.1016/j.fuel.2016.12.035.

CHAPTER 4

Thermal Plasma Waste Treatment

Milan Hrabovský

4.1 CHARACTERISTICS OF PLASMA PROCESSING

Among various waste treatment technologies, thermal plasma treatment offers specific performance characteristics. The energy delivered by plasma is used for melting and vitrification of inorganic materials and for gasification of organic substances. While decomposition of waste and dangerous materials in thermal plasmas has been intensively studied in the last decade (Heberlein and Murphy 2008; Gomez et al. 2009; Ruj and Ghosh 2014; Moustakas et al. 2005; Katou et al. 2001; Sakai and Hiaraoka 2000; Cheng et al. 2002; Rutberg et al. 2013; Chen et al. 1997; Poiroux and Rollin 1996), and industrial-scale systems for treatment of various types of waste have been installed (Westinghouse; PlascoEnergy Group; Pyrogenesis; CO. Tetronics; Solena; AlterNrg; Shuey and Ottmer 2006), a waste to energy process through plasma gasification of organics is a newly appearing application. The principal goal of the gasification is the production of fuel gases, principally the mixture of carbon monoxide and hydrogen, called syngas. Alternatively, the production of hydrogen together with solid carbon (carbon black, carbon nanoparticles) is a new application studied in recent years.

Thermal plasma enables the decomposition of organics by pure pyrolysis in the absence of oxygen, or with a sub-stoichiometric amount of added oxygen (gasification), leading to the production of high-quality syngas,

DOI: 10.1201/9781003096887-4

with high content of hydrogen and carbon monoxide and minimum presence of other components. In the gasification process, oxygen is supplied to balance carbon and oxygen molar concentrations to suppress the production of solid carbon and for the maximum production of CO and maximum heating value of produced gas. As the main goal of this technology is the production of syngas, the energy balance of the process and syngas quality are much more important than in the case of waste treatment, where the principal goal is material decomposition.

A number of non-plasma systems have been developed for the production of syngas from waste organics and biomass. The principal limitations of most of these methods are the low heating value of produced syngas and the production of tar formed from complex molecules of hydrocarbons created during the process at lower temperatures. The gas from low-temperature gasification typically contains only 50% of energy in syngas components CO and H_2, while the remainder is contained in CH_4 and higher aromatic hydrocarbons (Boerrigter 2005). The possibility of controlling syngas composition in most of these processes is limited. The need for the production of clean syngas with controlled composition results in the use of technologies based on external energy supply. Energy can be carried into the gasification reactors by hot gases or solids (sand), a relatively new method is based on the use of thermal plasmas.

Plasma pyrolysis and gasification leading to the production of syngas (Fabry et al. 2013; Lemmens et al. 2007; Tanga and Huanga 2005; Diaz et al. 2015; Luche et al. 2012; Hrabovsky et al. 2006; Hrabovsky et al. 2009; Hrabovsky et al. 2014; Hlina et al. 2014; Hrabovsky et al. 2011; Rutberg et al. 2004; Zasypkin and Nozdrenko 2001) is an alternative to non-plasma methods of organic waste and biomass treatment. Plasma is a medium with the highest content of energy; therefore, substantially lower plasma flow rates are needed to supply a sufficient amount of energy compared to other media used for this purpose. The result is minimum contamination and dilution of the produced syngas by plasma gas and an easy control of syngas composition. The process also acts as an energy storage—electrical energy is transferred into plasma enthalpy and then stored in the produced syngas. The main advantages are a better control over the composition of a syngas; its higher calorific value; and reduction of undesired contaminants like tar, CO_2, CH_4, and higher hydrocarbons. Another advantage of plasma is the wide choice of materials to be treated. As the energy needed for the process is supplied by plasma and chemical reactions are not the primary source of energy, the process can be applied to a wide range of organic materials. These advantages of plasma technology, together with its higher energy consumption, must be considered when evaluating the technical and economic feasibility of plasma treatment.

Plasma treatment offers a better control over the temperature of the process, higher process rates, lower reaction volume, and especially an optimum composition of produced syngas. The process exploits the thermochemical properties of plasma. Decomposition of the material is achieved by the action of the kinetic energy of plasma particles, which is extremely high due to high temperatures of the plasma. In addition, the presence of charged and excited species makes the plasma environment highly reactive, which can catalyse homogeneous as well as heterogeneous chemical reactions. The main advantage of plasma consists in much higher enthalpy and temperature of plasmas than those of gases used in non-plasma methods. Thus, substantially lower plasma flow rates are able to carry sufficient energy for the process, and the composition of the produced syngas is not much influenced by plasma gas composition. Moreover, substantially less energy is required to heat the plasma gas to the reaction temperature.

4.2 PROCESSES IN PLASMA REACTOR

Plasma treatment of waste materials exploits energy and thermochemical properties of thermal plasmas. Electric energy is transformed in a plasma generator into enthalpy of plasma. Plasma delivers energy needed for the destruction of materials supplied into a reactor volume where energy is transferred to the treated material. A principal scheme of plasma-material treatment process is presented in Figure 4.1. The treated material is supplied into the reactor where it interacts with plasma flow which is produced in plasma generators positioned outside of the reactor. Energy is transferred from flowing plasma to material and causes its heating and thermal destruction, melting, and volatilisation. Additional gases can be added into the reactor for the control of composition of reaction products. Gases produced by the volatilisation of organic components react with plasma gas under high temperature which is maintained in the reactor due to the input of energy carried by plasma. Gaseous reaction products flow into the quenching chamber where gases are rapidly cooled down to suppress chemical reactions at lower temperatures and preserve the gas composition corresponding to high temperatures. Melted inorganic components flow out of the reactor where they are cooled down and solidified.

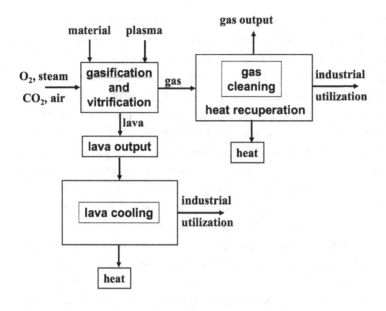

FIGURE 4.1 Principal scheme of processes of plasma waste treatment.

The role of plasma in the plasma waste treatment can be summarised as follows:

- Transport of energy, needed for the endothermic reaction of material decomposition, melting, and volatilisation; into the reactor volume.
- Energy transfer to treated material.
- Control of temperature inside the reactor volume.
- Supply of chemicals for the control of composition of reaction products.

Plasma is generated in plasma torches where thermal plasma is produced by ionisation of plasma gas in an electrical discharge. Thermal plasma for waste treatment is mostly generated in plasma torches with electric arcs either in a transferred or in a non-transferred mode. In the non-transferred mode, the arc is ignited inside the plasma torch, both electrodes are parts of the torch, and generated plasma exits the torch through an exit nozzle, which can be created by one of the electrodes. For transferred arcs, one of the electrodes is a material to be treated. The advantage of transferred arcs follows from high heat transfer to treated material and low power loss to the torch body, because the arc column is positioned inside the reactor volume and the material is in direct contact with the arc discharge. However, the usage of this type of arc is limited to the treatment of electrically conducting materials. A much wider range of applications exist for non-transferred arc plasma systems that can be applied for any kind of material including organics and biomass as well as for liquids or gases. Peak plasma temperatures at the exit of the plasma torch are typically from several thousands of K to more than 20,000 K. Treated material is in contact with the plasma jet exiting the torch attached to the reactor wall. The temperature inside the reactor can be controlled by arc power, material feed rate, and plasma gas flow rate.

Due to high temperatures in the reactor volume, the organic material components are volatilised and molecules of produced gas are dissociated into atoms and/or simpler molecules. Gases produced by material volatilisation react with plasma components, and the resulting composition of reaction products depends on the temperature in the reactor volume. The temperature is determined by an energy balance between an energy input carried by plasma flow, an energy spent for material heating, an energy spent or produced by chemical reactions, and an energy spent for melting inorganic components.

Gasification of organic materials involves several chemical reactions. High temperature inside the gasification reactor leads to heterogeneous gas-solid phase reactions. The following basic homogeneous gas–gas phase reactions take place in the reactor atmosphere in gases produced by the volatilisation of organics containing mostly hydrogen, oxygen, and chemically bound carbon.

Exothermic reactions:

Oxidation

$$C + O_2 \leftrightarrow CO_2 \quad \Delta H = -393.51 \, kJ/mol \quad (4.1)$$

$$H_2 + \tfrac{1}{2}O_2 \leftrightarrow H_2O \quad \Delta H = -241.83 \, kJ/mol, \quad (4.2)$$

$$CO + \tfrac{1}{2}O_2 \leftrightarrow CO_2 \quad \Delta H = -282.98 \, kJ/mol, \quad (4.3)$$

Partial oxidation

$$C + \tfrac{1}{2}O_2 \leftrightarrow CO \quad \Delta H = -110.53\,kJ/mol \qquad (4.4)$$

Water gas-shift reaction

$$CO + H_2O \leftrightarrow CO_2 + H_2 \quad \Delta H = -41.15\,kJ/mol. \qquad (4.5)$$

Methane production

$$CO + 3H_2 \leftrightarrow CH_4 + H_2O \quad \Delta H = -206.17\,kJ/mol. \qquad (4.6)$$

Endothermic reactions:

$$C + CO_2 \leftrightarrow 2CO \quad \Delta H = +172.45\,kJ/mol \qquad (4.7)$$

$$C + H_2O \leftrightarrow CO + H_2 \quad \Delta H = +131.30\,kJ/mol \qquad (4.8)$$

Steam and dry methane reforming

$$CH_4 + H_2O \leftrightarrow CO + 3H_2 \quad \Delta H = +206.17\,kJ/mol \qquad (4.9)$$

$$CH_4 + CO_2 \leftrightarrow 2CO + 2H_2 \quad \Delta H = +247.32\,kJ/mol \qquad (4.10)$$

Other chemical reactions producing more complex molecules can take place if the temperature in the reactor is not sufficiently high. Plasma may bring a number of other species, namely radicals, ionised atoms, and molecules. After mixing with gases, produced by volatilisation of treated material, a number of chemical reactions take place. High temperature and the presence of radicals and ions lead to an increase in reaction rate. Conditions in the reactor volume should lead to complete mixing of all components, and the resulting temperature in the volume must be high enough to ensure an appropriate composition of reaction products. In the optimal case, the conditions in the reaction zone of the reactor volume should be close to the state of thermodynamic equilibrium, and the final composition of the produced gas is then determined by the temperature and pressure.

4.3 PLASMA PYROLYSIS, GASIFICATION, AND PLASMA-AIDED COMBUSTION

The main objective of the gasification process is the production of syngas, i.e., gas mixture with prevailing contents of hydrogen and carbon monoxide and small admixtures like CH_4, CO_2, and other minor components. Principally all carbon and hydrogen atoms from treated organics can be used for syngas production if material and produced gases are heated to sufficiently high temperature. A maximum material to syngas conversion efficiency is achieved if all carbon is oxidised to CO. To prevent the production of solid carbon and ensure complete carbon gasification, an oxidising agent is added if carbon molar content in treated material is higher than oxygen content. This is usually done by the addition of oxygen, air, steam, or CO_2.

Depending on the amount of added oxygen, the following three processes are distinguished:

- Pyrolysis—no oxygen added—all energy for the endothermic reaction is supplied by plasma.

- Gasification—oxygen is added to balance the number of carbon and oxygen atoms.

- Combustion—oxygen is added for the production of CO_2 and H_2O and energy production by oxidation. The major part of energy needed for material decomposition is supplied by oxidation, the minor part of energy is brought by plasma.

The three processes are schematically shown in Figure 4.2.

The processes differ in the amount of externally added oxygen to the reaction volume. In plasma pyrolysis, only plasma and material are supplied into the reactor. All energy needed for destruction and volatilisation of material and for chemical reactions between produced gases and plasma gas is supplied by plasma. The heating of material leads to its volatilisation, high temperature in the reactor volume causes dissociation of molecules of gas produced by the volatilisation, and a gas is produced with CO and H_2 as main components. If the total number of carbon atoms supplied in treated material and plasma is higher than the number of oxygen atoms, then solid carbon is produced in the form of char or carbon black powder. In special conditions, mostly with the use of catalysers, carbon

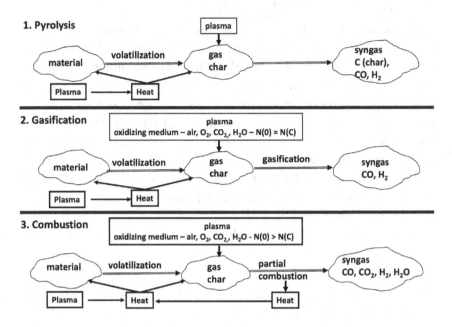

FIGURE 4.2 Plasma-assisted pyrolysis, gasification, and combustion.

nanoparticles are produced. If the amount of oxygen in all reactants is small, prevailing reaction products are hydrogen and solid carbon. Plasma pyrolysis of materials with low oxygen content is used for the production of carbon black and hydrogen. The maximum outputs of hydrogen and carbon are obtained by the pyrolysis of organics with no oxygen content like polyethylene with plasma without oxygen like methane.

In case of gasification, a certain amount of oxygen is supplied into the reactor to balance molar fractions of carbon and oxygen in input reagents. Oxygen molar fraction can be increased by the addition of air, oxygen, carbon dioxide, steam, or water. Most energy for the endothermic process of material gasification is supplied by plasma, and in case of the usage of air or oxygen, an additional energy comes from the oxidation of volatilised material. In this case, the process is similar to partial oxidation, but for most organic materials, including biomass, some energy must be added by plasma to achieve complete gasification. Oxygen can also be added in the form of carbon dioxide or steam. In these cases, additional energy is needed for their dissociation. In an ideal gasification process, all carbon atoms are bound in CO molecules and produced syngas consists of only carbon monoxide and hydrogen. The ratio of molar fractions of hydrogen to carbon monoxide, which is an important parameter for syngas utilisation, can be controlled by the combination of various oxidation agents.

The following three processes represent common gasification routes:

a) Gasification by the reaction with oxygen

$$M + \frac{(n_C - n_O)}{2} O_2 \Rightarrow n_C CO + n_{H_2} H_2. \qquad (4.11)$$

b) Gasification by the reaction with steam

$$M + (n_C - n_O) H_2O \Rightarrow n_C CO + (n_{H_2} + n_C - n_O) H_2. \qquad (4.12)$$

c) Gasification by the reaction with CO_2

$$M + (n_C - n_O) CO_2 \Rightarrow (2n_C - n_O) CO + n_{H_2} H_2, \qquad (4.13)$$

where $n_C = c/M_C$, $n_{H2} = h/2M_H$, and $n_O = o/M_O$ are molar concentrations of carbon, hydrogen, and oxygen in treated material M with mass fractions of carbon, hydrogen, and oxygen equal to c, h, and o, respectively.

The resulting composition of reaction products is determined by the composition of treated material, plasma gas, and added gases, as well as by conditions inside the reactor, especially by temperature. Thermal plasma supplied to a reactor is close to the state of thermodynamic equilibrium and thus the plasma properties and plasma composition are determined by temperature and pressure. The conditions for thermodynamic equilibrium should be established also in gases produced by the material treatment. This can be achieved if temperature in a reactor volume is sufficiently high. The temperature is determined by the balance between energy supplied by plasma and by exothermic chemical reactions in produced gases and the energy spent for material decomposition and endothermic reactions. The resulting composition of produced gases can be then determined by thermodynamic computations.

As an example of pyrolysis process, Figure 4.3 presents the temperature dependence of the equilibrium composition of the system containing carbon, hydrogen, and oxygen with mass fractions corresponding to a fir wood. The equilibrium composition of this heterogeneous

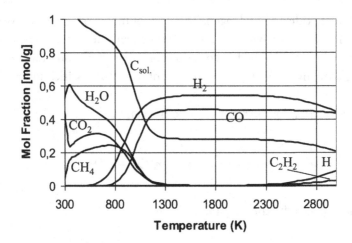

FIGURE 4.3 Composition of products of wood pyrolysis. The mass ratios of components in wood: $c = 0.511$, $h = 0.064$, and $o = 0.425$ (Hrabovsky et al. 2014).

system was calculated using the method described in (Coufal 1994), and the input data for calculations of standard reaction enthalpies and standard thermodynamic functions of system components were taken from database (Coufal et al. 2005). Molar fractions of components in the gas phase are shown for gas components H_2, H_2O, CO, CO_2, CH_4, and C_2H_2. For solid carbon (C_s), the ratio of the number of solid carbon moles to the number of all moles in the gas phase is given. It can be seen that the material is decomposed into hydrogen, carbon monoxide, and solid carbon with a small amount of other components at temperatures above 1,200 K.

The example of gasification process is shown in Figure 4.4, which presents the composition of products of gasification of the same material with an addition of CO_2 and oxygen in steam/argon plasma. The amount of added oxygen and carbon dioxide was determined to balance carbon and oxygen moles contained in all reacting components. It can be seen that the optimum composition of syngas with high concentrations of H_2 and CO and with the minimum presence of solid carbon and other gas molecules is obtained, similarly as in the case of pyrolysis, at temperatures higher than 1200 K. The minimum optimal value of temperature close to 1200 K can be found for gasification of all organic materials.

It was confirmed in experiments in real plasma reactors that the composition of plasma gasification products is close to the calculated equilibrium composition obtained in thermodynamic computations. Examples

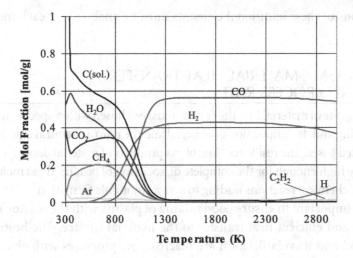

FIGURE 4.4 Composition of products of wood gasification. Wood 47 kg/h, humidity 6.5%, CO_2 115 slm, oxygen 30 slm, steam plasma 18 g/min, and argon 13.5 slm. The mass ratios of components in wood: $c = 0.511$, $h = 0.064$, and $o = 0.425$ (Hrabovsky et al. 2014).

of experiments with solid and liquid organic waste and gas hydrocarbons and comparison with results of calculations are presented in Section 4.6.

In case the number of oxygen moles n_O in reacting gases is higher than the number of carbon moles n_C, i.e., $n_O/n_C > 1$, a carbon dioxide and water molecules are formed and the reaction energy of oxidation is produced. For higher values of n_O/n_C, a major part of the energy for material decomposition is supplied by oxidation, and plasma brings only a minor part of the energy and its role is mainly ignition of combustion process, control of temperature, and supply of reacting gases. The process is used in industrial waste treatment systems with high material throughputs. These systems usually utilise high-power DC or AC arc torches operated with high flow rates of air. Additional air is often supplied into the reactor to ensure sufficient oxidation for the production of energy needed for material decomposition, melting, and gasification. The quality of the produced gas is lower due to the higher content of carbon dioxide and other products of combustion. The process is used in the systems for waste treatment with high throughputs of material, where destruction of waste is the principal goal of the technology and the quality and heating value of syngas is not so important.

In this chapter, only reactions of basic components of organic materials, i.e., carbon, hydrogen, and oxygen were considered. Since organic materials can contain also small amounts of other elements, possible chemical reactions of these additional elements must be analysed for each individual case.

4.4 PLASMA-MATERIAL HEAT TRANSFER AND PROCESS RATE

Feeding rate of material into the reactor must be chosen with respect to the rate of heating and decomposition of material and the rate of chemical reactions in produced gases. The residence time of gas products of material decomposition must be high enough for the complete dissociation of produced gas molecules and for chemical reactions leading to syngas molecule formation.

It is important to ensure good mixing of plasma with the reactor atmosphere and efficient heat transfer to the material surface. The heating of material and its volatilisation are macroscopic processes with characteristic times much higher than time constants of chemical reactions at high temperatures. Thus, the rate of gasification is determined dominantly by the rate of material heating and volatilisation.

The principal processes of heat transfer from plasma to material are convection and radiation. The basic relationship of heat transfer to the material is

$$q = \alpha P \left(T_p - T_s \right) + Q_r, \tag{4.14}$$

where q is the heat flux, P is the area of material surface, α is the heat transfer coefficient, T_p is the plasma temperature, T_s is the material surface temperature, and Q_r is the power transfer by radiation. Radiation transfer plays an important role in plasma temperatures above 10,000 K; for most cases, plasma temperature inside the reactor is lower due to a plasma jet expansion and an entrainment of surrounding gas, and heat transfer is dominated by convection. The heat transfer coefficient α is a complicated function of many quantities like plasma properties and characteristics of plasma flow. It is given by the relation

$$\alpha = \frac{k}{D} f \left(\text{Re}, \text{Pr}, \ldots \right), \tag{4.15}$$

where k is the thermal conductivity of plasma, D is the linear dimension of solid body, and f is a function of characteristics of plasma flow around the material, which are described by dimensionless criteria of the flow field like Reynolds number Re and Prandtl number Pr.

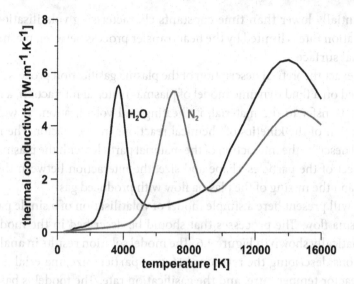

FIGURE 4.5 Dependence of thermal conductivity on temperature for plasmas of steam, nitrogen, and a mixture of argon and hydrogen (volume ratio argon/hydrogen is 33/10).

The important plasma property for heat transfer from plasma to material surface is thermal conductivity k. The dependence of thermal conductivity on temperature for three common plasma gases is shown in Figure 4.5.

It can be seen that steam plasma has the highest values of thermal conductivity for all temperatures. Moreover, the peak in thermal dependence of k, corresponding to the dissociation of steam molecules, is at relatively low temperatures below 4,000 K, which is close to temperatures in the surrounding of treated material. Due to its high thermal plasma conductivity, very high plasma enthalpy, and optimal plasma composition (hydrogen, oxygen) the steam plasma is an optimal choice for waste treatment technology. Unfortunately the above-mentioned steam plasma characteristics are also responsible for high erosion rates of electrodes in plasma sources with electric discharges.

The rate of plasma gasification process is dependent on material volatilisation rate, which is controlled by heat transfer to the material surface, and on the kinetics of chemical processes in the gas phase produced by volatilisation. A gas sheath is created between the material surface and plasma due to rapid material volatilisation. The sheath substantially reduces heat transfer to material and reduces the volatilisation rate. As time constants of chemical reactions in the gas phase at elevated temperatures are substantially lower than time constants characterising volatilisation, the gasification rate is limited by the heat transfer process between plasma and material surface.

An exact theoretical description of the plasma gasification process should be based on a fluid dynamic model of plasma–material interaction, a model of heat transfer to the material, its heating and volatilisation, as well as a description of the kinetics of chemical reactions in the reactor. The model should describe the interaction of the material particles with the plasma flow, the effect of the particles' shape and size, the interaction between the particles, and the mixing of the plasma flow with produced gas.

We will present here a simple model of volatilisation of a single particle in plasma flow. The processes that should be described in the model are schematically shown in Figure 4.6. The model solution results in analytical equations describing the relations between particle size, material properties, reactor temperature, and the gasification rate. The model is based on a solution of the Arrhenius equation describing volatilisation of material at a given temperature and equations describing heat and mass transfer between the reactor atmosphere and the surface of treated material.

The rate of material volatilisation is described using the Arrhenius equation:

$$\dot{m} = A \exp\left(-\frac{E}{RT_s}\right), \qquad (4.16)$$

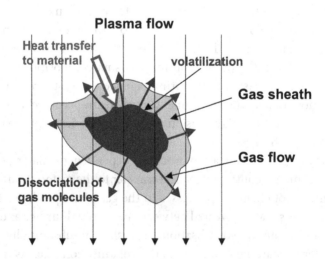

FIGURE 4.6 Material particle in plasma flow.

which determines the dependence of volatilisation rate $\dot{m}$ on the temperature of material surface T_s, A is the frequency factor, E the activation energy, and R is the universal gas constant.

To describe the heat transfer to the particle, we use the film model (Bird et al. 2002). Heat flux through the sheath created around a spherical particle with surface temperature T_s is given by the equation:

$$q_0 = \frac{\dot{m} C_p (T_r - T_s)}{e^{-\frac{\dot{m} C}{\alpha}} - 1}, \qquad (4.17)$$

where $\dot{m}$ is the volatilisation rate, T_r is the temperature in plasma flow outside the sheath, C_p is the specific heat, and α is the heat transfer coefficient, both corresponding to local conditions in the sheath. We will approximate the heat transfer coefficient by the relation for heat transfer to the sphere in a flowing fluid (Bird et al. 2002):

$$\alpha = \frac{k \cdot Nu}{D} = \frac{k}{D}\left(2 + 0.6 \operatorname{Re}^{\frac{1}{2}} \operatorname{Pr}^{\frac{1}{2}}\right), \qquad (4.18)$$

where Nu is the Nusselt number, Re is the Reynolds number and Pr is the Prandtl number, D is the diameter of the sphere, and k is the thermal conductivity of gas within the sheath.

The relation between the volatilisation rate and heat flux is given by the energy balance equation:

$$q_0 = \dot{m} \Delta h_{gas}, \qquad (4.19)$$

where Δh_{gas} is the energy needed for material gasification. It is supposed that all heat flux to the surface is spent for volatilisation.

From Equations (4.16)–(4.19), we get the following relation between the mass gasification rate $\dot{m}$ and the difference of temperatures in the reactor T_r and at the particle surface T_s:

$$\dot{m} = \frac{\alpha}{C_p} \ln\left(\frac{C_p}{\Delta h_{gas}}(T_r - T_s) - 1\right), \qquad (4.20)$$

By solving Equations (4.16)–(4.20), we obtain the dependence of the volatilisation rate and surface temperature T_s on the temperature in the reactor T_r and the particle diameter D.

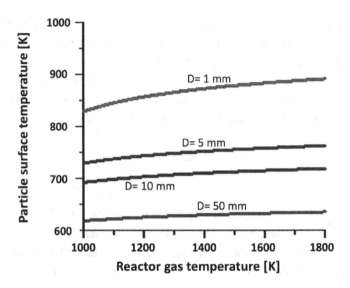

FIGURE 4.7 Surface temperature of wood particles as a function of reactor temperature for various wood particle diameters.

Dependences calculated for various diameters of spherical wood particles are presented in Figures 4.6–4.8. The computations were made for input wood characteristic parameters ($c = 0.511$, $h = 0.064$, $o = 0.425$, $A = 7.7 \cdot 10^6$ s^{-1}, $E = 1.11 \ 10^5$ J/mol), and sheath values of transport and thermodynamic coefficients k, h, C_p corresponding to a

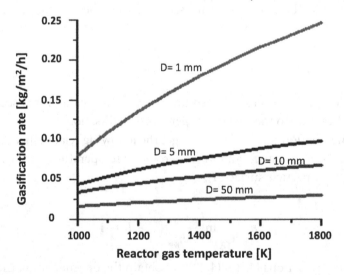

FIGURE 4.8 Gasification rate of wood particles as a function of reactor temperature for various wood particle diameters.

mixture of hydrogen and CO with the volume ratio of 1:1, for zero relative velocity between a particle and the surrounding gas, and for averaged sheath temperature

$$T_{sheath} = \frac{(T_r - T_s)}{2} \qquad (4.21)$$

Figures 4.7 and 4.8 present the effect of the reactor temperature on the particle surface temperature and on the gasification rate. It can be seen that both parameters are substantially influenced by the particle diameter. An increase in the diameter results in the reduction of heat transfer to the particle, as the particle is more efficiently shielded by the gas sheath formed from the volatilised material.

In Figure 4.9, the ratio of the volume of particles to their gasification rate is plotted as a function of the temperature. Relations between particle volume, size, and process rate should be taken into account when selecting an appropriate reactor volume, process temperature, and particle size for a specific material. A minimum reactor volume required for a given material throughput can be determined from these dependences. Reactor volume should be several times higher than the volume occupied by the particles. Figure 4.9 shows that the required reactor volume rapidly increases with particle size.

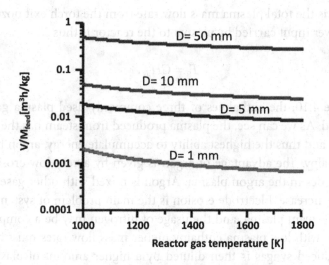

FIGURE 4.9 Ratio of volume of supplied wood particles to gasification rate.

4.5 ENERGY BALANCE OF PLASMA TREATMENT

The primary role of plasma in the waste treatment process is the transport of energy needed for material decomposition into the plasma reactor and the transfer of energy to treated material. Power delivered into the reactor by plasma is given by an integral of plasma enthalpy flow over the cross section of plasma torch exit nozzle

$$F_h = 2\pi \int_0^R \rho v h r \, dr, \tag{4.22}$$

where ρ is the plasma density, h the plasma enthalpy and v the flow velocity, and R is the radius of the torch exit nozzle. The total enthalpy flux in the exit nozzle of the torch is determined by the plasma torch power $W = I.U$ where I is the current and U is the arc voltage, and power loss P_{torch} to the torch body,

$$F_h = W - P_{torch}. \tag{4.23}$$

Plasma enthalpy in waste treatment systems is often characterised by mass-averaged enthalpy H_{av} which can be evaluated from easily measured data,

$$H_{av} = \frac{W - P_{torch}}{G}, \tag{4.24}$$

where G is the total plasma mass flow rate from the torch exit nozzle. The total power input carried by plasma to the reactor is thus

$$F_h = G H_{av}. \tag{4.25}$$

In Figure 4.10, the enthalpies of three commonly used plasma gases are compared. As we can see, the plasma produced from steam has the highest enthalpy and thus the highest ability to accumulate energy, argon has very low enthalpy. The advantage of argon is given by a very low erosion rate of electrodes in the argon plasma. Argon is mixed with other gases for an enthalpy increase. Electrode erosion is the main problem of systems operated with steam plasma, and the usage of nitrogen may be a compromise. For gases with low plasma enthalpy higher mass flow rates must be used, and produced syngas is then diluted by a higher amount of plasma gas. Steam plasma offers not only high plasma enthalpy but also high thermal

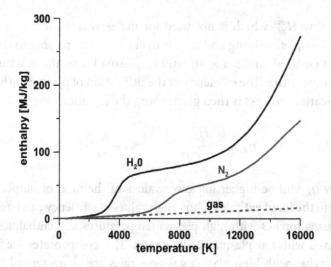

FIGURE 4.10 Dependence of plasma enthalpy on temperature for three plasma gases.

conductivity, and thus high heat transfer from the plasma to the material. Moreover, the composition of plasma (hydrogen and oxygen) does not contribute to the dilution of syngas by other gases, which is a problem especially with plasma torches operated with high flow rates of air or nitrogen.

Plasma gasification systems utilise mostly arc plasma torches (Tanga and Huanga 2005; Hrabovsky et al. 2006; Hrabovsky et al. 2009; Hrabovsky et al. 2014; Hlina et al. 2014; Hrabovsky et al. 2011; Rutberg et al. 2004; Zasypkin and Nozdrenko 2001), and the common level of power loss in arc plasma torches is 10% to 50% of electric power W. Generally, a higher efficiency and thus a lower power loss can be achieved in plasma torches with high plasma gas flow rates G as plasma temperatures are lower and the high flow rate of colder gas surrounding the high temperature arc column reduces heat transfer from the plasma to the arc chamber walls and electrodes. On the other hand, high plasma gas flow rates lead to a higher power needed for heating plasma gas to the temperature inside the reactor, which must be kept high enough to ensure proper conditions for chemical reactions. Plasma enthalpy available for material processing ΔH is given by the difference of input total plasma enthalpy H_{av}^{in} and plasma enthalpy H_{av}^{out} corresponding to the temperature of gases leaving the reactor

$$\Delta H = GH_{av}^{in} - GH_{av}^{out}. \tag{4.26}$$

The enthalpy H_{av}^{out} which is not used for material treatment can be even higher if complete mixing and heat transfer from the plasma to the material is not ensured, and some amount of plasma leaves the reactor with a higher temperature. The efficiency of the utilisation of plasma enthalpy for the gasification process is then given using the equation

$$\eta_R = \frac{G\left(H_{av}^{in} - H_{av}^{out}\right)}{GH_{av}^{in}} = 1 - \frac{H_{av}^{out}}{H_{av}^{in}}. \tag{4.27}$$

Efficiency η_R will be higher for lower values of the ratio of output plasma enthalpy to the input enthalpy. Thus, high values of efficiency can be achieved using plasma torches with high plasma temperatures and enthalpies, which are achieved with low plasma gas flow rates. This compensates the fact that plasma torches with high plasma gas flow rates are characterised by lower plasma torch losses. Total efficiency determined by both the torch efficiency and the reactor process efficiency (Equation 4.27) must be taken into account.

The energy balance of processes in the reactor volume should include material heating, melting, and volatilisation, and chemical reactions in produced gases, energy loss due to the reaction products outflow from the reactor, and energy loss into the reactor walls. Plasma interacts with treated material in the reactor volume, and the heat transferred to a material causes melting of inorganic and volatilisation of organic material components. Plasma is mixed with produced gases, and due to the high temperature in the reactor volume chemical reactions take place between gases produced by organics volatilisation, plasma gas, and added gases. These reactions can be exothermic or endothermic depending on the composition of reacting gases, especially on the amount of oxygen. As shown in Section 4.3, the temperature for the production of syngas with optimal composition with the maximum content of hydrogen and carbon monoxide and the minimum presence of other components should be higher than about 1200 K for most organic materials.

The power balance of processes in the reactor volume can be written as

$$GH_{av} = Q_{inorg}^{melt} + P_{inorg}^{out} + Q_{org}^{volat} + Q_{org}^{chem} + P_{org}^{out} + P_{reactor} + P_{plasma}, \tag{4.28}$$

where Q_{inorg}^{melt} is a power spent for inorganics melting, Q_{org}^{volat} a power for organics volatilisation, Q_{org}^{chem} a power needed for chemical reactions, P_{inorg}^{out}, P_{org}^{out} are a power losses carried out of the reactor by products of treatment of inorganics and organics, P_{plasma} a power of plasma not transferred to material, and $P_{reactor}$ a power loss to the reactor walls.

All terms on the right-hand side of (4.28) with the exception of Q_{org}^{chem} are always positive and represent power consumption. The term Q_{org}^{chem} corresponding to chemical reactions in the gas phase can be positive or negative depending on organics composition, plasma composition, and added gases. For plasma waste treatment systems with high material throughputs, a large amount of oxygen or air is added and energy is produced by partial combustion in the reactor volume, and the term Q_{org}^{chem} is then negative and represents power input. Plasma is in these cases used especially for the control of temperature distribution inside the reactor. The power input by plasma enthalpy creates only a small part of power spent for material treatment.

The temperature in the reactor volume is controlled by the energy balance (4.28). The temperature can be easily controlled by changing plasma power and the feed rates of material and added gases.

Energy balance has primary importance especially for applications where the production of syngas by plasma gasification is a main goal of the process. We will analyse the energy balance of plasma gasification of biomass (wood) by reactions with oxygen, steam and CO_2 described by Equations (4.11)–(4.13). All three reactions are for most organic materials endothermic. The energy needed for realisation of the reactions has to be supplied by plasma. It includes the energy needed for the production of gases by the volatilisation of material and the energy for reactions in a gas mixture of volatilised material, added oxidisers, and plasma. The evaluation of reaction enthalpy from the enthalpies of formation of reaction components is often impossible as the data are not known for most treated solid materials. The evaluation can be made on the basis of data about the heat of combustion of materials that are usually available.

The energy Δh_r needed for realisation of reactions represented by Equations (4.11)–(4.13) can be evaluated from known heat of combustion on the basis of scheme presented in Figure 4.11. The scheme corresponds to the gasification of biomass by reaction (4.11). The heat of gasification, i.e., production of syngas with composition n_c CO + n_{H2} H_2, is calculated as the difference of heat of combustion $\Delta h_{c,\,net}$ of biomass and heat of combustion of syngas $\Delta h_{c,\,syng}$

$$\Delta h_{gas} = \Delta h_{c,net} - \Delta h_{c,syng}. \qquad (4.29)$$

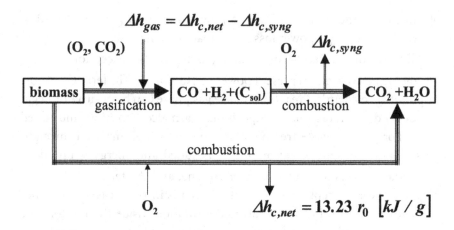

FIGURE 4.11 Scheme of reactions for the determination of reaction heat for biomass gasification.

The heat of combustion of syngas produced by the reaction (4.11) is given by using the equation

$$\Delta h_{c,\text{syng}} = n_{H_2} \Delta H_2 + n_C \Delta H_3, \qquad (4.30)$$

where $\Delta H_2 = -241.83$ kJ/mol is the enthalpy of reaction (4.2) and $\Delta H_3 = -282.98$ kJ/mol is the enthalpy of reaction (4.3). The heat of combustion $\Delta h_{c,\text{net}}$ is known for many materials used as fuels.

The heat of combustion of cellulosic materials, like wood or other biomass materials, can be calculated from the equation (Dietenberger 2002):

$$\Delta h_{c,\text{net}} = 13.23 \, r_0 \, [\text{kJ/g}], \qquad (4.31)$$

where r_0 is an external oxygen mass fraction needed for complete combustion:

$$r_0 = (8/3)c + 8h - o. \qquad (4.32)$$

The heat of combustion of syngas (low heating values, LHVs) produced by complete gasification of biomass by reaction (4.11) can be expressed as

$$\Delta h_{c,\text{syng}} = n_C \left(\Delta_f H^\circ(CO_2) - \Delta_f H^\circ(CO) \right) + n_{H_2} \Delta_f H^\circ(H_2O), \qquad (4.33)$$

where $\Delta_f H^\circ$ are the heats of formation of individual molar components.

For the reactions (4.12) and (4.13), the reaction heats include also heat of dissociation of H_2O and CO_2, respectively. The corresponding values of $\Delta h_{c,\text{syng}}$ can be calculated according to (4.33) where n_{H_2} and n_C are numbers of moles of H_2 and CO in syngas given by the equations (4.12) and (4.13).

The gasification reactions (4.11) to (4.13) can be realised only at elevated temperature T_r.

The total heat needed for gasification is then given by the sum

$$\Delta h_r = \Delta h_{\text{gas}} + \Delta H_T, \qquad (4.34)$$

where Δh_{gas} is given by (4.29) and ΔH_T is the heat needed for heating of all components on the right-hand side of Equations (4.11)–(4.13) from standard temperature to the reaction temperature T_r.

Figure 4.12 shows components of energy balance for gasification of a biomass (wood), calculated by solving Equations (4.29)–(4.34) for process

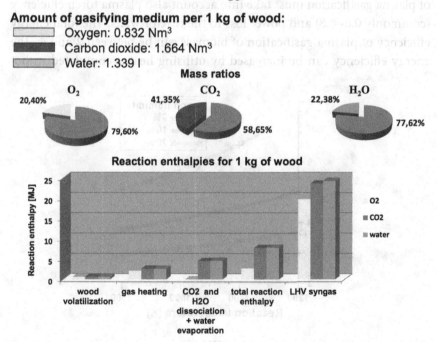

FIGURE 4.12 Mass and heat balances for the process of dry wood gasification by reactions with oxygen, water, and carbon dioxide. Mass ratios of components in wood: $c = 0.511$, $h = 0.064$, and $o = 0.425$.

of reaction of wood with oxygen (Equation 4.11) and for reactions with steam (4.12) and with CO_2 (4.13). The amounts of oxidising agents needed for complete gasification are shown for the three processes characterised in (4.11)–(4.13). Figure 4.12 presents the energy needed for volatilisation of wood and its gasification $\Delta h_{c,net}$, energy for heating of all components in the reactor to 1,200 K, energy for dissociation of added steam and carbon dioxide, and the sum of all reaction enthalpies. Low heating values (LHV) of produced syngas, i.e., heating values without the heat produced by steam condensation, are shown in the last columns in Figure 4.12.

An effect of a choice of oxidising agent on an energy efficiency is illustrated in Figure 4.12. If oxygen is used, the ratio of syngas heating value to the total enthalpy needed for realisation of the process is 7.7. If water or carbon dioxide are used, this ratio is lower than one half of this value.

The ratio of low heating value of syngas (LHV) to the energy needed for its production in dependence on reaction temperature is plotted in Figure 4.13 for three values of wood humidity for gasification by reaction with oxygen, and in Figure 4.14 for CO_2 process. Total energy efficiency of plasma gasification must take into account also plasma torch efficiency (commonly 0.6–0.8) and power loss to the reactor walls. The total energy efficiency of plasma gasification of biomass can be between 2 and 4. The energy efficiency can be increased by utilising heat accumulated in hot

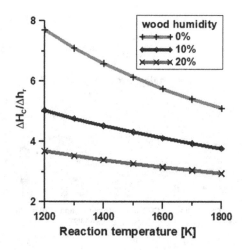

FIGURE 4.13 Dependence of ratio of LHV of syngas on total reaction enthalpy for gasification of wood by reaction with oxygen on reaction temperature for three values of wood humidity. Mass ratios of components in wood: $c = 0.511$, $h = 0.064$, and $o = 0,425$.

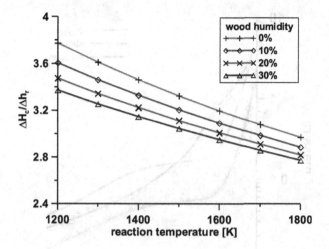

FIGURE 4.14 Dependence of ratio of LHV of syngas on total reaction enthalpy for gasification of wood by reaction with carbon dioxide on reaction temperature for three values of wood humidity.

syngas. As all power losses are losses to the cooling water of the torch and the reactor, the real power gain after recuperation of heat in a cooling system could be higher.

Mass ratios of components in wood: $c = 0.511$, $h = 0.064$, and $o = 0,425$.

The ratio of heating value of produced syngas to the energy needed for its production can be increased if the amount of oxygen added is higher than needed for balancing carbon and oxygen molar concentrations. Besides hydrogen and carbon monoxide, also carbon dioxide and steam are produced and reaction enthalpy is reduced due to additional oxidation. The LHV of produced syngas is lowered for the same value. The ratio of syngas heating value to needed energy input is thus increased. In large plasma waste treatment and gasification systems oxygen is often added through an introduction of a large amount of air. Air is supplied into the reactor also in the form of air plasma if air torches are used. Syngas heating value is then reduced due to the presence of CO_2 produced by partial combustion and also due to presence of nitrogen. However, the ratio of output to input energy is increased.

A substantial advantage of plasma treatment is in reduction of mass flow rate of gasifying medium compared to the flow rate of gases used for non-plasma gasification. Thus, in case of plasma gasification, the produced syngas is less diluted by gas supplied into the reactor and has higher heating value. Also the power losses connected with the heating of the

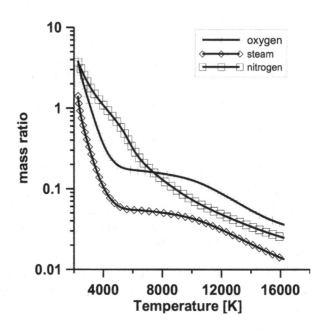

FIGURE 4.15 Ratio of mass of gas carrying energy Δh_r for complete gasification of wood, to the mass of wood, in dependence on gas temperature.

gasifying medium on the reaction temperature are reduced. The ratio of mass of plasma, or gas at lower temperatures, needed for supply of energy Δh_r for complete gasification, to the mass of treated material, is plotted in dependence on a gasifying medium temperature in Figure 4.15 for nitrogen, oxygen, and steam plasmas. The curves were calculated from thermodynamic equilibrium enthalpies of three gases (Boulos et al. 1994) and from the total energy of gasification determined above. For temperatures lower than 3000 K, the ratio is close to 1. Thus, for gasification with hot air (the amount of needed gas will be close to the values for nitrogen), almost half of the weight of produced syngas is air and thus syngas is diluted by a high percentage of nitrogen (approximately 39% of weight of produced syngas). For comparison with plasma systems: for steam plasma with input temperature 16 000 K, this ratio is less than 0.02 and thus almost undiluted syngas is produced.

4.6 GASIFICATION OF VARIOUS ORGANICS IN STEAM PLASMA

Steam and CO_2 plasmas can be used for production of pure syngas with high content of hydrogen and CO and minimum other components. Examples of syngas composition obtained by gasification of several solid and liquid organic materials in steam plasma are given in Table 4.1. All materials were treated under the same conditions in a reactor with inner volume 0.2 m^3 (Hrabovsky et al. 2006; Hrabovsky et al. 2009; Hrabovsky et al. 2014; Hlina et al. 2014; Hrabovsky et al. 2011). The power of the plasma torch was between 110 kW and 140 kW. The torch efficiency was 65%, thus the total plasma enthalpy flow into the reactor volume was 72 to 91 kJ/s. The mass flow rate of steam plasma was 18 g/min and the temperature measured at the reactor walls was 1100–1300 K. The following materials were treated:

- Fir wood saw dust (humidity 12%).
- Wooden pellets 5 mm diameter, 10 mm long (humidity 7.0%).
- Sunflower seeds skins.
- Soft brown coal (lignite) powder (humidity 45%).
- Polyethylene pellets, diameter 3 mm.
- Shredded waste plastics from PE bottles, 2–10 mm particles.
- RDF (refuse-derived fuel) processed from waste excavated from landfill sites. Composed of municipal solid waste (59%) and industrial waste (41%).
- Oil produced by a low-temperature pyrolysis of used tires. (>21 wt% of water, LHV 39.5 MJ/kg, molecular composition C_5H_8O). It contained a number of complex hydrocarbons including harmful and dangerous ones.

Various combinations of oxidising media (water, CO_2, O_2) were added. The ratio of molar concentrations C/O was close to 1 for all cases. No additional oxidising agent was added in case of lignite with high humidity, as water content in material supplied enough oxygen to ensure carbon/oxygen balance.

Material feed rates and flow rates of added oxidisers are given in Table 4.1. Values of carbon yield (C_{yield}) given in the table correspond to the ratio of the carbon content in product gases to carbon content in all input reagents.

It can be seen that syngas with high concentrations of H_2 and CO was produced for all materials. The third component with concentration

TABLE 4.1 Input Reagents, Composition of Syngas, and Its Heating Values

	Inputs						Syngas				
Material	Feed Rate (kg/h)	CO$_2$ (slm)	O$_2$ (slm)	H$_2$O (g/min)	C$_{yield}$	H$_2$ (%)	CO (%)	CO$_2$ (%)	CH$_4$ (%)	O$_2$ (%)	LHV (MJ/m^3)
Wood	41	-	64	18	1.0	45	39	15	1	0	10.1
Wood	41	125	-	18	0.9	42	42	15	1	0	10.2
Wood	60	86	66	18	1.0	41	52	5	1	1	11.1
Wood pellets	60	248	-	18	0.8	42	53	4	0	1	11.2
Seed skins	95	120	30	18		76	15	3	5	0	10.0
Coal	60	-	-	18		61	25	13	1	0	10.1
Polyethylene	11	210	80	18	1.0	35	42	22	0	1	9.1
Plastics	11	300	-	18	0.7	42	50	7	0	1	10.8
RDF	30	216	118	18	0.83	30	48	20	2	0	10.0
RDF	30	-	-	403	0.85	56	30	10	4	0	11.3
Pyrolysis oil	11	-	89	18	1.0	45	48	2	1	4	11.2
Pyrolysis oil	22	200	89	18	0.8	34	44	16	3	3	10.1

higher than several percent was CO_2. Concentrations of CH_4 and O_2 were very low in all cases. A certain small amount of solid carbon was produced in all experiments. This could be reduced by addition of more oxidising admixtures. For all materials, the content of tar and higher hydrocarbons in the product gas was substantially below 10 mg/Nm³. This content is lower than in most of non-plasma gasifiers, where the tar content for various types of reactors varies from 10 mg/Nm³ to 100 g/Nm³.

Composition of the syngas was close to the one determined by thermodynamic equilibrium computations as illustrated for gasification of wood and wooden pellets in Figure 4.16 and Figure 4.17. Measured syngas composition is compared with the composition calculated from Equation

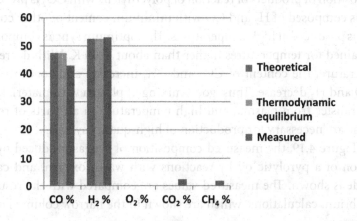

FIGURE 4.16 Composition of syngas produced by saw dust gasification (60 kg/h, water 159 g/min). Comparison of measured concentrations with the results of thermodynamic equilibrium calculations and from Equation (4.12) (theoretical).

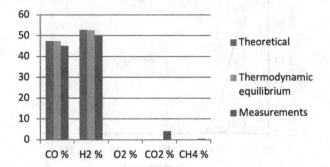

FIGURE 4.17 Calculated and measured composition of syngas. Wooden pellets (10–15 mm)—30 kg/h, water 79.6 g/min, steam plasma 18 g/min. Theoretical—Equation (14.12).

(4.12) and composition obtained from thermodynamic computations for the state of thermodynamic equilibrium at $T = 1200$ K.

The composition calculated by thermodynamic calculations and the composition determined from Equation (4.12) are almost the same, measured data are close to calculated values, experimental values of CH_4 and O_2 concentrations were lower than 1%, about 4% of CO_2 was present in experimentally determined composition.

In some experiments presented in Table 4.1, the content of CO_2 was higher than the one predicted by equilibrium calculations for $T = 1200$ k. The content of carbon dioxide and methane in syngas increases with decreasing temperature. This can be seen in Figure 4.18 where equilibrium composition of products of reaction of polyethylene with CO_2 is presented. Syngas composed of H_2 and CO with minimum content of other components is produced at high temperatures. This optimum syngas composition is obtained for temperatures higher than about 1200 K. With decreasing temperature, the content of CO_2 and CH_4 increases and concentrations of CO and H_2 decrease. Thus, good mixing of plasma with material, high heat transfer from plasma, and high temperature in all parts of reactor volume are necessary for production of high-quality syngas.

In Figure 4.19, the measured composition of syngas produced by gasification of a pyrolytic oil by reactions with water, oxygen, and carbon dioxide is shown. The measured values are compared with the results of equilibrium calculations, which are shown in the narrow columns on the

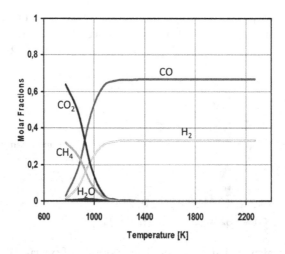

FIGURE 4.18 Thermodynamic equilibrium composition of products of polyethylene gasification (reaction of 1 mole of CO_2 with 14 g of polyethylene).

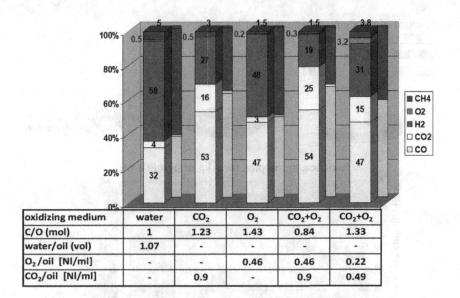

FIGURE 4.19 Measured composition of syngas produced by gasification of a pyrolytic oil by reactions with water, oxygen, and carbon dioxide. Calculated equilibrium values are shown in the narrow columns on the right-hand side of measured results.

right-hand side of measured results. It can be seen that under some conditions, the measured syngas composition is close to the equilibrium one.

Mass and energy balances of gasification of pyrolytic oil by oxygen, steam, and CO_2 processes for molar ratio O/C = 1 are shown in Figure 4.20. The following equations describe the oil gasification processes:

$$C_5H_8O + 2O_2 \rightarrow 5CO + 4H_2 \qquad (4.35)$$

$$C_5H_8O + 4H_2O \rightarrow 5CO + 8H_2 \qquad (4.36)$$

$$C_5H_8O + 4CO_2 \rightarrow 9CO + 4H_2 \qquad (4.37)$$

The total energy balance of gasification of biomass by reactions with oxygen and carbon dioxide is given in Table 4.2. The values for gasification of 1 kg of wood are given. The first two lines give the values of energy needed for chemical processes and for heating of all reagents to the temperature 1200 K. The lines in the bottom of the Table 4.2 show the values of heat of combustion of wood, heating value (LHV) of produced syngas, ratio of syngas LHV to energy spent for the process, and ratio of syngas LHV

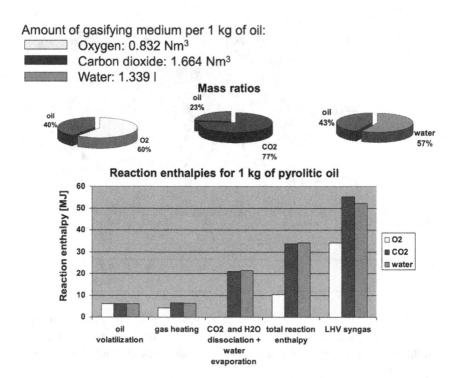

FIGURE 4.20 Mass and energy balances of gasification of 1 kg of pyrolytic oil.

to the total energy input (sum of energy spent for the process and heat of combustion of wood).

If water or CO_2 are used as oxidising medium, the ratio of heating value of syngas to energy needed for syngas production is lower than in the oxygen process, as additional energy is needed for water and CO_2 dissociation. Lower energy supplied by plasma is needed in the oxygen process if a

TABLE 4.2 Energy Balance of Wood Gasification

Energy Consumption for 1 kg of Wood (MJ)				
Oxidising gas	Reaction $co_2 \to CO + O$	Reaction $wood + O \to CO + H_2$	Heating to $T=1$ 200 K	Total energy consumption
O_2	-	0.61	1.97	2.58
CO_2	4.6	0.61	2.39	7.6
Energy Balance of 1 kg Wood Gasification				
Oxidising gas	Wood (Heat of combustion) (MJ)	Syngas (low heating value) (MJ)	$\dfrac{\text{Syngas LHV}}{\text{Energy consum.}}$	$\dfrac{\text{Syngas LHV}}{\text{Energy input.}}$
O_2	19.2	20	7.8	0.92
CO_2	19.2	24.6	3.2	0.92

higher amount of oxygen (O/C > 1) is added. Part of the carbon is oxidised to CO_2 and additional energy is produced by this oxidation. The LHV of produced syngas is then reduced for the same energy due to an increase in CO_2 concentration. The process with high surplus of oxygen is used in large gasification units for higher material throughputs to reduce power requirements of plasma generators. The total energy efficiency, i.e. the ratio of heating value of syngas to total energy input (the sum of energy consumed for the process and the heating value of material) is almost the same for both processes, as a decrease of needed energy input is compensated by a decrease of the heating value of syngas.

REFERENCES

AlterNrg, http://alternrg.com/.
Bird, R.B., W.E. Stewart, and E.N. Lightfoot. 2002. *Transport Phenomena*. New York: John Willey & Sons, Inc.
Boerrigter, H., and B. van der Drift. 2005. ""Biosyngas" Key-Intermediate for Production of Renewable Transportation Fuels, Chemicals and Electricity." ECN report ECN-RX-05-181. In *14th European Biomass Conference & Exhibition*, Paris.
Boulos, M., P. Fauchais, and E. Pfender. 1994. *Thermal Plasma Fundamentals and Applications*. New York: Plenum Press.
Chen, X., J.M. Badie, and G. Flamant. 1997. "Dynamics of Complex Chemical System Vaporization at High Temperature. Application to the Vitrification of Fly Ashes by Thermal Plasma." *Chemical Engineering Science* 52: 4381–4391.
Cheng, T.W., J.P. Chu, C.C. Tzeng, and Y.S. Chen. 2002. "Treatment and Recycling of Incinerated Ash Using Thermal Plasma Technology." *Waste Management* 22: 485–490.
CO. TETRONICS, http://www.tetronics.com/.
Coufal, O. 1994. "Composition of the Reacting Mixture SF6 and Cu in the Range from 298.15 to 3000 K and 0.1 to 2 Mpa." *High Temperature Materials and Processes* 3: 117–139.
Coufal, O., P. Sezemsky, and O. Zivny. 2005. "Database System of Thermodynamic Properties of Individual Substances at High Temperatures." *Journal of Physics D: Applied Physics* 38: 1265–1274.
Diaz, G., N. Sharma, E. Leal-Quiros, and A. Munoz-Hernandez. 2015. "Enhanced Hydrogen Production Using Steam Plasma Processing of Biomass: Experimental Apparatus and Procedure." *International Journal of Hydrogen Energy* 40: 2091–2098.
Dietenberger, M. 2002. "Update for Combustion Properties of Wood Components." *Fire and Materials* 26: 255–267.
Fabry, F., Ch. Rehmet, V. Rohani, and L. Fulcheri. 2013. "Waste Gasification by Thermal Plasma: A Review." *Waste Biomass Valor* 4: 421–439.

Gomez, E., D. Amutha Rani, C.R. Cheeseman, D. Deegan, M. Wise, and A.R. Boccaccini. 2009. "Thermal Plasma Technology for the Treatment of Wastes: A Critical Review." *Journal of Hazardous Materials* 161: 614–626.

Heberlein, J., and A.B. Murphy. 2008. "Thermal Plasma Waste Treatment." *Journal of Physics D: Applied Physics* 41: 053001.

Hlina, M., M. Hrabovsky, T. Kavka, and M. Konrad. 2014. "Production of High Quality Syngas from Argon/Water Plasma Gasification of Biomass and Waste." *Waste Management* 34, no. 1: 63–66.

Hrabovsky, M., M. Hlina, M. Konrad, V. Kopecky, O. Chumak, A. Maslani, T. Kavka, O. Zivny, and G. Pellet. 2014. "Steam Plasma-Assisted Gasification of Organic Waste by Reactions with Water, CO_2 and O_2." In *21st International Symposium on Plasma Chemistry (ISPC 21)*, Cairns.

Hrabovsky, M., M. Hlina, M. Konrad, V. Kopecky, T. Kavka, O. Chumak, and A. Maslani. 2009. "Thermal Plasma Gasification of Biomass for Fuel Gas Production." *Journal of High Temperature Material Processes* 13: 299–313.

Hrabovsky, M., M. Konrad, V. Kopecky, and M. Hlina. 2006. "Pyrolysis of Wood in Arc Plasma for Syngas Production." *Journal of High Temperature Material Processes* 10, no. 4: 557–570.

Hrabovsky, M., M. Konrad, V. Kopecky, M. Hlina, T. Kavka, O. Chumak, and A. Maslani. 2011. "Steam Plasma Gasification of Pyrolitic Oil from Used Tires." In *Proceedings of 20th International Symposium on Plasma Chemistry*, Philadelphia, PA.

Katou, K., T. Asou, Y. Kurauchi, and R. Sameshima. 2001. "Melting Municipal Solid Wasteincineration Residue by Plasmamelting Furnace with a Graphite Electrode." *ThinSolid Films* 386: 183–188.

Lemmens, B., H. Elslander, I. Vanderreydt, K. Peys, L. Diels, M. Oosterlinck, and M. Joos. 2007. "Assessment of Plasma Gasification of High Caloric Waste Streams." *Waste Management* 27: 1562–1569.

Luche, J., Q. Falcoz, T. Bastien, J.P. Leninger, K. Arabi, O. Aubry, A. Khacef, J.M. Cormier, and J. Lédé. 2012. "Plasma Treatments and Biomass Gasification." *IOP Conference Series: Materials Science and Engineering* 29: 012011.

Moustakas, K., D. Fatta, S. Malamis, K. Haralambous, and M. Loizidou. 2005. "Demonstration Plasma Gasification/Vitrification System for Effective Hazardous Waste Treatment." *Journal of Hazardous Materials* 123: 120–126.

PlascoEnergy Group, http://www.plascoenergygroup.com/.

Poiroux, R., and M. Rollin. 1996. "High Temperature Treatment of Waste: From Laboratories to the Industrial Stage" *Pure and Applied Chemistry* 68: 1035–1040.

Pyrogenesis, http://www.pyrogenesis.com/.

Ruj, B., and S. Ghosh. 2014. "Technological Aspects for Thermal Plasma Treatment of Municipal Solid Waste—A Review." *Fuel Processing Technology* 126: 298–308.

Rutberg, P.G., A.N. Bratsev, and A.A. Ufimtsev. 2004. "Plasmochemical Technologies for Processing of Hydrocarbonic Raw Material with Syngas Production." *Journal of High Temperature Material Process* 8, no. 3: 433–446.

Rutberg, P.G., V.A. Kuznetsov, E.O. Serba, S.D. Popov, A.V. Surov, G.V. Nakonechny, and A.V. Nikonov. 2013. "Novel Three-Phase Steam–Air Plasma Torch for Gasification of High-Caloric Waste." *Applied Energy* 108: 505–514.

Sakai, S., and M. Hiaraoka. 2000. "Municipal Solid Waste Incinerator Residue Recycling by Thermal Processes." *Waste Management* 20: 249–258.

Solena, http://www.solenagroup.com/.

Shuey, M.W., and P.P. Ottmer. 2006. "LLW Processing and Operational Experience Using a Plasma Arc Centrifugal Treatment (PACT™) System." In WM'06 Conference, February 26–March 2, Tucson, AZ.

Tanga, L., and H. Huanga. 2005. "Biomass Gasification Using Capacitively Coupled RF Plasma Technology." *Fuel* 84: 2055–2063.

Westinghouse, http://www.westinghouse-plasma.com/projects/.

Zasypkin, L.M., and G.V. Nozdrenko. 2001. "Production of Acetylene and Synthesis Gas from Coal by Plasma Chemical Methods." In *Thermal Plasma Torches and Technologies*, Vol II., ed. O.P. Solonenko. Cambridge: Interscience Publishing, 234–243.

CHAPTER 5

Product Applications

Guido Van Oost

5.1 INTRODUCTION

Since the 1980s, the applications of thermal plasmas have increased significantly. In the 1990s, basic research led to major advances in understanding the fundamental phenomena involved, and to a renewed interest in the use of thermal plasmas in materials processing and waste disposal (Heberlein and Murphy 2008). The use of plasma torches for environmental purposes is a relatively new process.

Thermal plasma pyrolysis and vitrification is a double simultaneous disintegration process. Thermal plasmas (produced in plasma torches) offer an alternative and superior solution for the treatment of waste streams. The feedstock is treated with thermal plasma in a reactor chamber, where organic components are converted into a calorific syngas (a mixture of hydrogen and carbon monoxide), while the inorganic components can be converted into a slag directly in the process. Due to the unique property of strongly intensifying the energy content of the process gas, plasma torches offer very clear advantages over traditional combustion which relies on the energy content of the waste as a heat source. In the plasma torch, the process heat is supplied directly by heat transfer through the electric arc discharge. The use of electrical energy also enables the allothermic mode of operation and reduces the required oxygen supply (which is normally injected as air along with a lot of nitrogen, increasing the volume of the waste gas to be treated), and allows better control of the chemical processes. Another advantage is the ability to produce a syngas free of

DOI: 10.1201/9781003096887-5

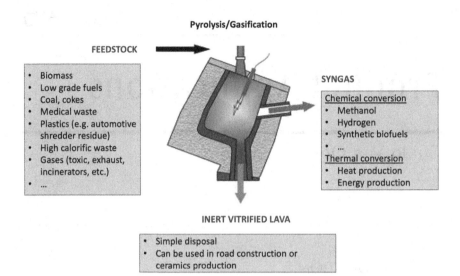

FIGURE 5.1 Potential applications of thermal plasma technology.

nitrogen by providing exactly the amount of gasifier (e.g., CO_2) to allow for carbon volatilisation or solid carbon production.

The treatment of raw materials is optimised with regard to the quality of the syngas and the recovery process according to criteria determined by the end user. These criteria can be the maximum energy content of the syngas for combined electricity/heat production (*thermal conversion*), or *chemical conversion* for the production or recovery of valuable products from the syngas such as hydrogen, methanol, synthetic fuel feedstock, ammonia, olefins, or other liquid hydrocarbons via Fischer–Tropsch synthesis, as schematically shown in Figure 5.1.

Increasingly strict legislation on the treatment of waste streams and the limitations of conventional technologies such as thermal combustion, catalytic oxidation, and adsorption make plasma technologies increasingly attractive. *The driving force is to prioritise environmental quality at affordable costs and to contribute to sustainable development*

The use of thermal plasmas has the following advantages:

- Much higher temperatures can be achieved than with conventional heat generators.
- Highly reactive (reactive species such as atomic oxygen and hydrogen) environment and reducing atmosphere in the gasification process resulting in lower NO*x* emissions.

- High energy density and high heat transfer efficiency, enabling shorter residence times and high throughput.
- Because sufficiently high temperatures and homogeneous temperature distribution can be easily maintained throughout the reactor volume, the production of higher hydrocarbons, tar, and other complex molecules is significantly reduced compared to combustion.
- Incomplete combustion gasification at high temperatures and rapid cooling of the syngas makes it possible to avoid the formation of dioxins and furans, the most dangerous toxic substances.
- Deep decomposition of waste into simple compounds, greatly simplifying cleaning against harmful impurities.
- Possibility of joint processing of different types of waste without pre-sorting, which is especially important for the processing of unsorted biomedical and other toxic waste.
- Possibility of processing difficult waste such as tyres, carpets, and sludge.
- Low thermal inertia and easy feedback control. Ability to quickly adapt the process by changing the flow rate of the oxidiser (air, steam, or other plasma-forming gas) and the power of the plasma torches. Ability to create a desired gas atmosphere. In addition, the low inertia of the process minimises the risk of harmful emissions in the event of an emergency, with the ability to quickly stop the process.
- Lower plasma gas input per unit of heating power than the gas flow of a conventional burner and thus lower energy loss corresponding to the energy required for heating gas to reaction temperature. Furthermore, the amount of gases diluting the syngas produced is lower.
- Significant reduction of the flue gas volume and thus the load on the gas cleaning system.
- Less entrainment of dispersed particles.
- Smaller installations than incinerators due to high energy densities, lower gas flows, and volume reduction.
- The heat source is electricity rather than the energy released during combustion and thus is independent of the treated materials,

providing flexibility, rapid process control, and more options in process chemistry, including the ability to generate valuable by-products. The temperature in the reactor can be easily controlled by changing plasma power and feed rate of material and added gases.

- Optimal control of the composition of the end-product in stable form. Ability to obtain more calorific and cleaner syngas from the organic part of the waste, which is not contaminated with by-products typical of conventional gasification (especially tar).
- Production of vitrified slag that can be used as building material.
- Vitrification of combustion ash.
- Production of value-added products (metals) from slag.

5.2 SINGLE VERSUS TWO-STAGE PLASMA GASIFICATION

A hybrid process including plasma treatment in combination with incineration or another thermal process can provide better utilisation of the calorific content of the waste.

The shortcoming of the in-flight single-stage plasma gasification system in converting the inorganic fraction of the treated material and immobilising the shaft fraction to a valuable vitrified by-product has been addressed through two-stage gasification systems in which the melting of the ash and metals is separated from the gasification process. Two-stage plasma gasification publicly available results are very scarce.

An interesting plasma gasification plant, well-documented in the literature, is the demonstration factory of *Advanced Plasma Power* (APP) in Swindon (UK). This two-stage system consists of an oxy-steam bubbling fluidised-bed gasification process step followed by an auxiliary plasma converter, in which the syngas and solid residue from the first stage are treated due to a long residence time at high temperatures (Materazzi et al. 2015). The heat required for the gasification reactions is produced by the complete oxidation of part of the *refuse-derived fuel* (RDF). Refuse is a general term for municipal solid waste (MSW) and industrial waste. The term RFD usually refers to the separated high calorific fraction of MSW.

A clear advantage of the two-stage system is the successful vitrification of the residual inorganic fraction leaving the plasma converter in a slag, while in the current single-stage in-flight plasma gasification system, the melting point of the ungassed fraction was not reached and most of this fraction was recovered as solid particles and thus has little or no value as a by-product of the gasification.

5.3 APPLICATIONS

Some potential applications are schematically shown in Figure 5.1.

Examples of thermal plasma treatment of different types of waste (municipal solid waste, biomass, hazardous waste, and sludge) will be given from the personal experience of the authors (collaboration between IPP-CAS Prague, Ghent University, Academy of Sciences of Belarus and an industrial partner).

5.3.1 Municipal Solid Waste

Municipal solid waste (MSW) accounts for a small share—about 5% of the total amount of waste. But it should get the most attention because here is the greatest number of problems concentrated, which are caused by the most rapid growth in volume, uncertainty of composition, release of the dangerous ingredients, pollution of soil, groundwater and atmosphere, and accumulation of huge amounts in landfills and illegal landfills, as well as regular landfill fires. Recently, the problem of the spread of infectious diseases through rodents, birds, and insects that have become accustomed to landfills has become particularly acute.

On the other hand, MSW contains many products and substances that can be disposed of for reuse. This is especially relevant for the recovery of energy (WTE), which is contained in significant amounts in the combustible part of the waste. Note that in many countries, MSW and other combustible waste are classified as renewable energy sources (RES), or more strictly defined as secondary RES. Plasma-assisted gasification in the WTE process combines the use of thermal plasma with partial oxidation of the hydrocarbons in MSW.

The International Energy Agency calls the management of energy waste with controlled high-temperature incineration and pollution control technology the best alternative to landfilling MSW.

The extremely complicated chemical composition of MSW, as well as their enormous diversity in size and phase state of fractions, makes it possible to formulate a fundamental premise: there is no single technology capable of processing waste under acceptable conditions.

The most promising method for processing both solid and liquid combustible waste is plasma gasification (Sikarwar et al. 2020; Agon 2016). Plasma methods have been used successfully in industry for decades. There are not many examples of commercial-scale plasma gasification, but there are a huge number of laboratory studies and pilot plants whose authors are unanimous in their opinion on the prospects of plasma gasification of combustible waste and the uniqueness of plasma technologies.

A contemporary approach is the use of automated waste sorting using intelligent robotic systems developed based on deep learning artificial neural networks. Such systems can recognise hundreds of thousands of objects and exceed human capabilities. It is even more advantageous if the installation will receive already sorted waste by introducing a system of separate collection from the population and organisations.

A popular method of using MSW for power generation is the pre-production of so-called refused-derived fuel (RDF), which consists of 2–3 cm pressed granules, produced from about one-third of the MSW remaining after sorting. In calorific value, 1.7 kg RDF corresponds approximately to 1 m^3 natural gas. The vast majority of MSW consists of combustible material. Energy recovery from waste is widespread in the world and is a global trend called *Waste-to-Energy*.

This approach is possible due to the high carbon content of MSW (and several other industrial wastes) and high heating value. The lower heating value (LHV) of MSW lies within the limits of 4.2–12.6 MJ/kg with an average value of 8.4 MJ/kg. In comparison, this lignite indicator varies in the range of 6.3–17 MJ/kg, i.e., it has a comparable value, so MSW can be considered a low-grade fuel. For rubber and plastic, these values reach 31.1 and 27.4 MJ/kg, respectively, which is two to four times higher than the corresponding values for lignite. Rubber and plastics are therefore considered to be the most suitable for power generation, especially when using gasification methods.

In terms of toxicity, dioxins and furans occupy a special place, which are formed from waste containing chlorinated derivatives (polyvinyl chloride, cardboard, newspapers, etc.) combustion. Dioxins and furans, which are formed both in landfills and uncontrolled incineration, are particularly dangerous. Once in the soil, dioxin is taken up by plants (especially through their underground part), soil fauna, transferring it to birds and other animals through the food chain. In addition, dioxin, which is carried away from the soil by air and water currents to the water bodies, also gets to birds and mammals via zooplankton, crustaceans, and fish. In other words, within vegetables, meat, and especially dairy and fish products taken from the contaminated area, dioxin will somehow end up on a human table. The high stability of this poison promotes repeated circulation through the food chain.

The proven approach to reducing the formation of dioxins is to provide zones with a high temperature above 1,200°C with a residence time of at least 2 s, when the dioxins have been completely destroyed, and after rapid cooling or catalytic afterburning to prevent a new process of dioxin formation.

5.3.2 Biomass

Experimental results have been obtained in the medium-scale thermal plasma gasification reactor equipped with a hybrid (gas-water) plasma torch with arc power up to 160 kW at the Czech Academy of Sciences (IPP-CAS) in Prague (Van Oost et al. 2009).

The gasification of biomass to synthesis gas (syngas) offers an alternative to fossil fuels. Conventional biomass gasification technologies are based on the reaction between a solid or liquid carbonaceous material (containing mainly chemically bonded carbon, hydrogen, and oxygen) and limited amounts of air or oxygen. The exothermic reactions provide sufficient energy to produce a primary gaseous product containing mainly CO, H_2, CO_2, and $H_2O(g)$ and a small amount of higher hydrocarbons. Usually, some heat is supplied to the reactor from external sources to control the process, but most of the heat comes from the calorific value of the biomass. The main problem with the common biomass gasification is the production of tar.

Thermal plasma offers the possibility to decompose biomass by pure pyrolysis in the absence of oxygen. In this process, all the energy required for gasification comes from the plasma, and no energy for decomposition is produced by combustion. The main advantage is better control of the composition of the produced gas; higher heat capacity of the gas; and reduction of unwanted contaminants such as tar, CO_2, and higher hydrocarbons. Most plasma gasification/pyrolysis experiments have been performed using arc plasma torches with relatively high plasma gas flow rates. The high flow rate of plasma ensures good mixing of plasma with the treated material and an even temperature in the reactor. However, the produced syngas contains plasma gas components, usually nitrogen and oxygen in case air or nitrogen plasma gases are used. The use of mixtures of inert gas with hydrogen overcomes this drawback but increases costs. Therefore, steam is used as plasma gas in experiments at IPP Prague.

5.3.3 Hazardous Materials

The most used treatment of *toxic organic waste* is thermal processing. Common practice is, for example, direct combustion in industrial furnaces and boilers. This usually concerns liquid and solid waste with moderate and high calorific values and a minimal content of halogens. However, the combustion conditions in these furnaces and boilers do not always correspond to the parameters necessary for the complete combustion of organic waste, leading to large emissions of harmful substances into the atmosphere. As a result, exhaust gases can contain dangerous products of incomplete chemical combustion. This is because the neutralisation process of organic waste by thermal methods is carried out at temperatures prone to the formation of other harmful compounds.

An alternative method of processing toxic organic waste is based on thermal plasma technology. The use of arc plasma with average temperatures in the order of 5000 K makes it possible to effectively carry out the destruction of organic compounds in atoms and ions at very high rates and a high degree of conversion. In addition, the destruction of complex compounds in the plasma is very efficient and can take place in the absence of oxygen, which provides the opportunity to successfully carry out plasma pyrolysis reactions, which in some cases has advantages over combustion.

In the framework of a *NATO for Peace Project*, at the A.V. Luikov Heat and Mass Transfer Institute of the National Academy of Sciences of Belarus in Minsk, a three-beam plasma reactor for the *destruction of perennial pesticides* was developed and built with a total power of 200 kW, in collaboration with Ghent University and IPP-CAS Prague (Van Oost et al. 2013). This reactor can use plasma-forming gases of different compositions and is easily tunable for the processing of different types of toxic organic waste, including waste belonging to the group of persistent organic pollutants (POPs). The heart of the plasma chemical reactor is a three-jet mixing chamber equipped with three arc plasma torches (power 50 kW; thermal efficiency up to 70%; plasma-forming gas air; plasma-forming gas consumption 3–6 g/s). The processing is characterised by a very turbulent plasma flow, which is formed in the three-jet mixing chamber, which guarantees a very intensive mixing of the plasma flow and the substances to be treated. The high temperature in the reactor leads to complete degassing of the inorganic ash residue. A shock cooling (quenching) module is used to prevent the formation of secondary toxic products. Acids are neutralised in the alkaline environment of a wet filter. An ion-mobility spectrometer RAID S2 is used for the detection of chemical agents in the air to control the exhaust gas composition and to prevent environmental disasters.

The tests showed that the thermal efficiency of the plasma torch in the temperature range of 3000–5000 K is 60–70%. By using different compositions of plasma-forming gas, the parameters of the medium in the reactor can be controlled to achieve a high level of toxic substance processing.

5.3.4 Sludge

An attractive application is the use of high-temperature plasma technology for processing problematic agricultural organic waste streams such as sludge, and the valorisation of the resulting syngas and slag/ash to maximise the extraction of phosphorus and high control of harmful components, to comply with the philosophy of the circular economy.

Sludge from sewage treatment is a waste material and is also considered a legal issue. This includes the need to treat sewage sludge according to the waste management hierarchy, where material and energy consumption are higher than simple disposal (usually landfill). In line with the principles of the circular economy, this is mainly achieved through their direct application to agricultural land (banned in some countries) or through composting, with subsequent use of this compost on agricultural land or for re-cultivation. From 2020, however, the microbiological criteria for the application of sludge in the soil are tightened and it is likely that landfilling will be banned completely in the European Union (EU) from 2024 (Moško 2022).

Municipal sewage sludge is rich in both organic matter and phosphorus. On the other hand, it is full of organic pollutants (e.g., hormones, chemotherapeutics, etc.) and heavy metals. Phosphorus is on the list of critical EU raw materials as an important nutrient and in some European countries it is already legally enforceable that phosphorus must be obtained from sewage sludge. At present, however, this acquisition is hampered by the economic barrier presented by the high price (cost) of recovered phosphorus, which cannot compete with the price of the primary feedstock. However, the latter is gradually increasing, which can be seen as an opportunity to store the ash from the combustion of sludge for the purpose of later phosphorus acquisition.

5.4 EXAMPLES OF PLASMA GASIFICATION PLANTS AT AN INDUSTRIAL SCALE

Plasma methods have been used successfully in the industry for decades. Plasma technology has long been used for surface coating and for the destruction of hazardous materials. There are not many examples of commercial-scale plasma gasification, but there are a huge number of laboratory studies and pilot plants whose authors are unanimous in their opinion on the prospects of plasma gasification of combustible waste and the uniqueness of plasma technologies.

Publicly available information on industrial-scale commercial plasma gasification plants is often very scarce (Sikarwar et al. 2020; Agon 2016). Therefore, most of the available data on these installations is based on their reported design criteria, while actual operating results may differ slightly from these figures.

A fairly large number of plasma gasification projects that were started up have failed or have been discontinued (see some examples below), not only due to financial, organisational, and planning errors (e.g., obtaining the necessary environmental permits), but also due to a number of technological bottlenecks that in some cases proved to be insurmountable. Moreover, the availability of these installations is lower than for incineration (8000 h/year). Many providers of this relatively new technology only have a pilot-scale installation that can process several tons per day, which leads to an over-ambitious scale-up as the capacity of the proposed plasma treatment systems is often up to two orders of magnitude greater. This can cause cold spots leading to incomplete gasification. Other potential technological problems include refractory lining material, high-temperature corrosion, and particle build-up and clogging of piping and afterburner. Adding these challenges to the technological barriers inherent in plasma torches (such as torch efficiency and electrode life), it becomes clear that a plasma gasification project must be very well designed.

An important element for the success of a plasma treatment plant project is the location, which is crucial to ensure the constant availability of waste for the plant. Therefore, the development of a plasma treatment facility is often a collaboration between the technology providing company, local private, and/or intermunicipal (waste) syndicates and companies that provide a stable market for the products of the process.

There are only a limited number of industrial-scale plasma waste treatment plants that have been in operation for several years, but the number of plasma waste treatment plants is growing all over the world due to their high destruction efficiency and environmentally friendly nature. The configuration of these installations differs in the type of gasifier, the type of thermal plasma used (AC or DC, transferred or not, and water-stabilised or gas-stabilised torches), the plasma unit (single or two-stage systems), the treated material, and the oxidising medium.

Currently, most of the high-power plasma torch market is shared by four companies namely Westinghouse, Europlasma, Tetronics, and Phoenix Solutions Company (Sikarwar et al. 2020). Phoenix, Europlasma, and Westinghouse use DC torches (both transferred and non-transferred) with water-cooled metal electrodes, while Tetronics uses transferred DC torches with two graphite electrodes and in collaboration with Advanced Plasma Power develops and commercialises thermal plasma-assisted waste-to-power plants based on the transferred DC technology. Westinghouse and Europlasma have established plasma waste-to-energy processes based on their own technologies. They have also developed market-ready plants through subsidiaries.

A schematic of a plasma gasification plant is shown in Figure 5.2, taking example by the treatment of sewage sludge.

For illustration, some examples are given below for different types of waste streams sorted according to the plasma technology supplier (Westinghouse Plasma Corporation, Europlasma, Tetronics, and Phoenix Solutions) (Agon 2016).

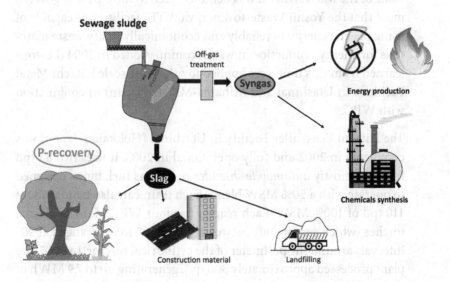

FIGURE 5.2 Schematic of plasma gasification plant for the valorisation of sewage sludge.

5.4.1 Westinghouse Plasma Corporation

Westinghouse Plasma Corporation (WPC) has a long track record in the plasma industry. They went from their knowledge of plasma torches to the development of plasma gasification systems. Their gasifiers are equipped with non-transferred plasma torches to provide some of the heat. The remainder of the energy required for waste treatment is obtained from the chemical energy of the waste and typically by the addition of coke (about 4% of the total charge).

WPC's Waltz Mill *plasma gasification pilot plant* is in Madison (Pennsylvania, USA) and has tested a wide variety of material feedstocks. This reactor was built in 1984 with a design capacity of 48 tons per day (tpd) and serves as a test unit for new project developments. Furthermore, WPC has several commercially operating reference projects with plasma-assisted gasifiers using WPC plasma torch technology:

- One of the first installations was the pilot-scale plasma gasification plant in Yoshii (Japan), operated by Hitachi Metals Ltd. It was commissioned in 1999 and was capable of processing 24 tons per day (tpd) of unprocessed *municipal solid waste* (MSW). The plasma energy supplied to this Plasma Direct Melting Reactor (PDMR) was 100 kWh/ton MSW, which is reportedly about 5% of the heating value of the MSW. After it was demonstrated to the Japanese government that the Yoshii Waste-to-Energy (WTE) facility was capable of using plasma energy to reliably and economically gasify waste materials for energy production, it was decommissioned in 2004. Lessons learned from this unit were applied to other full-scale Hitachi Metal Facilities in Utashinai and Mihama-Mikata (Japan) in conjunction with WPC.

- The Hitachi Eco-Valley Facility in Utashinai (Hokkaido, Japan) was completed in 2002 and fully operational in 2003. It uses two 83 tpd trains of mostly *automobile shredder residue* as fuel, but is designed to operate with a 50% MSW blend. Each train can also handle about 110 tpd of 100% MSW. Each reactor has four WPC Marc 3a plasma torches (which can operate between 80 and 300 kW) arranged at 90° intervals around the perimeter of the cylindrical reactor. In 2007, the plant processed approximately 300 tpd, generating up to 7.9 MWh of electricity and returning approximately 4.3 MWh to the electricity grid. Although successful, the plant is said to have closed in 2013, not because of technical problems, but because the increasing recycling of MSW in Japan has eliminated sufficient raw material for the plant's operation.

- The Hitachi Combined *MSW and Sewage Sludge* Gasification Plant, located between the two cities of Mihama-Mikata (Japan), processes 17.2 tons of MSW and 4.8 tons of sewage sludge and became operational in 2002. The solid waste is shredded on-site and the wastewater sludge is partially dried to 50% moisture content by the heat released by the combustion of the produced syngas. Due to its smaller capacity, the reactor has only two plasma torches arranged opposite each other. The slag produced (at 1.5 tpd) is used locally in road construction.

- In Ranjangaon, Pune (India), WPC has partnered with SMS Envocare Ltd (a group company of SMS Infrastructure Ltd.) to build the Maharashtra Enviro Power Ltd. (MEPL) factory, which has been in operation ever since 2009. The plant can process up to 72 tons of *hazardous waste* (with a calorific value of 3,000 MJm^{-3}), generating 16 MW of net electricity. The gasifier consists of two stages, with a coke bed reactor (at 1500°C) as the first stage and a second chamber for oxidation reactions at 1100°C. Four 300 kW plasma torches are installed, two of which operate continuously and the other two operate for an hour for lance (*removal of molten slag*), which is performed twice in an eight-hour shift. A similar 72 tons of hazardous waste for energy supply (named Vidharbha Enviro Protection Limited) is also being built in Mandwa, Nagpur (India), but no further information about this plant has been made available.

- The WPC technology is also applied in an *integrated biorefinery* operated by Coskata in Madison (USA). In this system, the plasma gasifier converts biomass and waste into syngas at the front and Coskata's syngas-to-biofuel conversion process at the back produces bioethanol.

- In Hubei (China), the Chinese energy company Wuhan Kaidi converts the syngas produced from plasma gasification with *biomass waste* as raw material into liquid fuels via their patented Fischer–Tropsch process. This demonstration facility has a 150 tpd WPC gasifier and was commissioned in 2013.

5.4.2 Europlasma

Europlasma provides industrial plasma-based solutions for waste recycling. Since 1992, through its subsidiary *Inertam*, it has been operating a plasma *vitrification plant for the conditioning of materials containing asbestos* in Morcenx (France) with a capacity of 8,000 tons per year (tpy). In 2014, a new plasma gasification installation next to this was put into operation by *CHO Power*. It is designed to convert 37,000 tpy of local industrial waste and 15,000 tpy of wood chips into the patented gasification technology, Turboplasma. The produced syngas generates a gross 12 MW of electricity and 18 MW of hot water that is valorised on-site in the wood dryer. The gasifier includes two plasma torches: the first to refine the raw synthesis gas produced during gasification, and the second to vitrify metals and minerals. *Europlasma Industries* treats hazardous waste.

5.4.3 Tetronics

One company that has published reliable experimental results of its plasma gasification demonstration unit is *Advanced Plasma Power Ltd* (APP). The company was founded in 2005 to commercialise the proprietary Gasplasma Technology originally developed by *Tetronics International* and demonstrated in its first test facility in Farringdon (UK). Since 2007, APP has been based in Swindon (UK), where they have conducted several plasma gasification experiments with municipal solid waste, mined landfill RDF, contaminated wood, and automobile shredder residue with a maximum design material feed rate of 100 kg/h. The technology is based on a separated oxy-steam gasification process followed by gas treatment by a DC plasma torch and vitrification of the inorganic material into a vitrified slag under the trademark PlasmaRok.

5.4.4 Phoenix Solutions

A plasma arc centrifugal furnace was developed in 1991 by Retech Systems LLC (Buffalo, NY, USA) at the Component Development and Integration Facility of the US Department of Energy in Butte (Montana, USA). The technology includes a centrifugal reactor volume equipped with two non-transferred plasma torches of 0.3 MW and 1.2 MW by *Phoenix Solutions Co*, Plymouth, Minnesota. The feed rate during trials with *metallic soils* at this site was approximately 55 kg/h. In Muttenz (Switzerland) a complete facility for the treatment of *hazardous and toxic waste* was operational in 1991 using Retech's plasma centrifugal furnace technology. This

PLASMOX facility, designed to operate at a feed rate of 1 ton per hour (tph), was owned and operated by MGC-Plasma AG (subsidiary of Burns and RoeGroup). In 1994, MGC-Plasma AG, in collaboration with Retech, also developed a PLASMOX portable unit model RIF2 that can be moved by four standard tractor-trailers. Under a contract from the Naval Research Laboratory, Retech designed and built the Plasma Arc Hazardous Waste Treatment System (PAHWTS) at the US Naval Base in Norfolk. Successful tests were performed with soil, epoxy paint, latex paint, a mixture of methanol with 1,1,1-trichloroethane (TCA) at 135–225 kg/h and oily rags at 91 kg/h. This plasma treatment system was sold through a government auction in 2007.

5.5 ECONOMICAL ASPECTS

The economic feasibility and efficiency of implementing a waste management system strongly depend on the composition of the processed waste and the region.

Although the costs associated with the installation and operation of plasma plants are very high on account of expensive equipment, demand for large amounts of electricity, and the necessity of highly skilled technicians and labour, the successful operations of various plants around the globe reflect positive economics (Xin-gang et al. 2016; Zhao et al. 2016).

The tipping fees combined with value-added by-products and finished products improve the cost-effectiveness of plasma treatment. The downside is the use of the most useful form of energy, namely electrical energy. However, this limitation can be overcome by making the plant self-sufficient in energy and reducing operating costs through recycling and utilisation of the energy content of the materials. Plasma treatment plants can be more economical than a conventional incinerator, provided energy and value-added products are recovered.

The main advantages of thermal plasma-assisted waste transformation processes over other processes are relatively less exhaust, higher volume reduction, installations with smaller footprints, faster start-up and shutdown times, and lower costs for a specified throughput.

The fact that thermal plasma technology is characterised by high energy consumption has hindered its further development for industrial applications. As environmental legislation becomes more stringent, the overall assessment of the technical and economic feasibility of thermal plasma technology must take into account its significant environmental benefits when compared to non-plasma methods.

In addition to electricity consumption, the main drawback of plasma technologies is the low life of the electrodes—from 100 to 300 h. Nevertheless, for example, replacement electrodes manufactured by Westinghouse Plasma Corporation can run for more than 1,000 h on average, and there is no problem replacing them (within 30 min without stopping the process).

In the past, the electrical energy required for plasma torches has hindered the further development of thermal plasma technology for industrial applications. Therefore, research into ways to increase the efficiency of plasma-based treatment processes is very important. On the other hand, with the massive deployment of renewable electricity sources, such as wind and solar, there is an opportunity to use cheap electricity generated during periods of low demand and high production—as such plasma gasification can be used to convert excess renewable energy into electricity into absorbable chemical energy. As environmental legislation becomes more stringent, the overall environmental and technical feasibility of plasma thermal technology must take into account its significant environmental benefits compared to non-plasma methods.

In the book of Kalogirou 2019 a WTE tool/business model is described. It pre-calculates and designs different types of WTE installations worldwide, taking into account multiple parameters such as the composition and calorific value of the input waste, local climate conditions, gross domestic product (GDP), social/financial conditions, and so on. An indicative range of initial capital expenditure (CAPEX) for the construction of a WTE plant is 450–900 USD per ton of annual capacity (tpa). Annual operating costs (OPEX) are influenced by several factors. A safe range to use in the event of a lack of sufficient data is from 2% to 3% of the facility's capital investment. As an example of thermal plasma WTEs, an installation in South Korea is mentioned for 10 tpd (3,300 tpa) MWS with a CAPEX of US$3 million.

From the scarce published literature and private communications, we give two more examples of thermal plasma WTE's.

1) *China*:

> Researchers (Li et al. 2016) investigated the process economics for an MSW plasma gasification plant with a treatment capacity of 600 tpd. The total investment amounted to US$99.3 million while the profits gained consisted of electricity, glass slag, foam glass, and

subsidy. They assumed the sales price of electricity to be 9.7 cents/kWh and sold about 350 kWh/t. So the revenue was about $6.67 million per year. The vitreous and foam slags were produced at a rate of 90,000 tpy and 10,000 tpy, respectively. The former was sold at US$105/t, while the latter was sold at US$1080.8/t, representing US$9.45 million and US$10.81 million per year, respectively. The factory was believed to receive a grant of $9.26, which amounts to $1.85 million. For example, total revenue generation for a year was approximately $28.87 million. According to this study, the economic efficiency is almost 3 times higher than with traditional combustion plants.

2) *Saudi Arabia* (Galaly and Van Oost 2017): 100 tpd of mixed solid waste produced during the pilgrimage season in Mecca. The initial estimate of CAPEX is US$37 million. The electrical power generation system required for the plasma treatment process was estimated to be 5,000 kW (2,000 kW used to operate the system and 3,000 kW sold), taking into account that: (1) the processor capacity is 100 tons of solid waste per day; (2) the sale of electricity amounts to 23.8 MW at a unit cost of the sale of electricity €0.18/kWh; (3) the profit from the sale of electricity is estimated at €4.36 million/year and the estimated profit from the treatment of solid waste is €6 million/year; and (4) the gross profit per ton of solid waste totalling €8 million/ year.

These examples indicate that the CAPEX for thermal plasma WTEs is within the range mentioned by Kalogirou, but on the higher side as expected. To fully assess the cost of a thermal plasma waste treatment facility, the cost of the facility, including waste gas treatment, the cost of energy required, flexibility in waste composition, government economic incentives, related regulations with the handling/processing/recycling of hazardous waste must be taken into account. The lack of published literature makes it impossible to draw firm conclusions about how a more efficient conversion of the heating value from waste to fuel gas or electricity would affect the economy of the plasma-assisted processes compared to conventional combustion processes.

Thermal plasma is used all over the world to recover metal from metallic waste such as red mud, electronic waste, waste circuit boards, electroplating sludge, aluminium slag, burnt ash, and so on. Various countries

(USA, Canada, China, Japan, Australia, Brazil, India, etc.) have conducted studies based on thermal plasma technologies to treat these wastes and recover the valuable metals. Most studies have used DC-transferred and non-transferred arc plasma systems. Studies on the recovery of metals from metallic waste are still in their infancy and there is a lack of data to set up large-scale facilities.

5.6 ENVIRONMENTAL ASPECTS

All materials can be decomposed if they are brought into contact with thermal plasma. The electrical energy from the flares goes to the plasma which transfers its energy to the substances to be treated, activating a two-fold simultaneous reaction process in the plasma chemical reactor: the organic compounds are thermally broken down into their constituent elements (mainly a mixture of carbon monoxide and hydrogen, called syngas with a more complete conversion of C to gas than in incinerators) and the inorganic materials are melted and converted to a dense, inert, non-leachable *vitrified slag* that does not require controlled disposal. Therefore, it can be seen as a completely closed treatment system. The main purpose of plasma treatment of organic waste is to produce syngas while the main purpose of combustion is material decomposition.

The main advantages of thermal plasma-assisted waste transformation processes over other processes are relatively less exhaust, greater volume reduction, smaller footprint installations, faster start-up and shutdown times, and lower costs for a specified throughput.

There is a lack of publicly available data from commercial installations. In an interesting and valuable study conducted by Herva and Roca 2013, the impact of different MSW treatment pathways was assessed based on environmental footprint and multi-criteria analysis. They used numerous environmental and sustainability indicators in their research. In addition, they emphasised integrative frameworks to obtain a more comprehensive assessment.

Their objectives include identification of the most advantageous waste treatment route from an environmental point of view and comparison of the results obtained from two different methods (ecological footprint as a single composite indicator and multi-criteria analysis integrating ecological footprint with other material flow indicators). They compared four different routes for processing MSWs, namely thermal plasma gasification, incineration with energy recovery, biological processing of organic fraction with energy recovery from RDF, and landfill. The results obtained in both cases proved that thermal plasma gasification is the best method to treat the waste, followed by biological treatment, incineration, and landfill (as the worst route). It should be noted that the ranking obtained was in accordance with the general hierarchy recommended by the public authorities favouring waste treatment technologies with energy and/or materials recovery.

Evangelisti and collaborators 2015 compared the environmental impacts of three different advanced MSW processing routes (gasification followed by plasma gas cleaning, rapid pyrolysis followed by incineration, and gasification with syngas combustion) with traditional technologies (landfill with electricity generation and incineration followed by electricity production). They showed that the environmental efficiency of two-stage plasma gasification was significantly higher than the conventional waste treatment routes and slightly better than an advanced incinerator (in Lincolnshire in the UK).

REFERENCES

Agon, N. 2016. *Development and Study of Different Numerical Plasma Jet Models and Experimental Study of Plasma Gasification of Waste*. PhD Thesis, Ghent University.

Evangelisti, S., C. Tagliaferri, R. Clift, P. Lettieri, R. Taylor, and C. Chapman. 2015. Life Cycle Assessment of Conventional and Two-Stage Advanced Energy-from-Waste Technologies for Municipal Solid Waste Treatment." *Journal of Cleaner Production* 100: 212–223.

Galaly, A.R., and G. Van Oost. 2017. "Environmental and Economic Vision of Plasma Treatment of Waste in Makkah." *Plasma Science and Technology* 19: 105503.

Heberlein, J., and A.B. Murphy. 2008. "Thermal Plasma Waste Treatment." *Journal of Physics D: Applied Physics* 41: 053001.

Herva, M., and E. Roca. 2013. "Ranking of Alternatives for the Treatment of Municipal Solid Waste Based on Ecological Footprint and Multi-Criteria Analysis." *Ecological Indicators* 25: 77–84.

Kalogirou, E.N. 2019. *Waste-to-Energy Technologies and Global Applications*. CRC Press and Taylor & Francis Group, Boca-Raton. https://www.google.it/books/edition/Waste_to_Energy_Technologies_and_Global/-f0wD-wAAQBAJ?hl=en&gbpv=1&printsec=frontcover.

Li, J., K. Liu, S. Yan, Y. Li, and D. Han. 2016. "Application of Thermal Plasma Technology for Solid Waste Treatment in China: An Overview." *Waste Management* 58: 260–269.

Materazzi, M., P. Lettieri, L. Mazzei, R. Taylor, and C. Chapman. 2015. "Fate and Behavior of Inorganic Constituents of RDF in a Two Stage Fluid Bed-plasma Gasification Plant." *Fuel* 150: 473–485.

Moško, J. 2022. *Modern Methods for Material and Energy Recovery from Sewage Sludge*. PhD Thesis, University of Chemistry and Technology in Prague and Ghent University.

Sikarwar, V.S., M. Hrabovský, G. Van Oost, M. Pohořelý, and M. Jeremiáš. 2020. "Progress in Waste Utilization via Thermal Plasma." *Progress in Energy and Combustion Science* 81: 100873.

Van Oost, G., et al. 2009. "Pyrolysis/Gasification of Biomass for Synthetic Fuel Production Using a Hybrid Gas-Water Stabilized Plasma Torch." *Vacuum* 83: 209–212.

Van Oost, G., M. Hrabovský, V. Kopecký, M. Konrád, M. Hlína, and T. Kavka. 2013. "Destruction of Toxic Organic Compounds in a Plasma-Chemical Reactor." *Vacuum* 88: 165–168.

Xin-gang, Z., J. Gui-wu, L. Ang, and L. Yun. 2016. "Technology, Cost and Performance of Waste-to-Energy Incineration Industry in China." *Renewable and Sustainable Energy Reviews* 55: 115–130.

Zhao, X., G. Jiang, A. Li, and L. Wang. 2016. "Economic Analysis of Waste-to-Energy Industry in China." *Waste Management* 48: 604–618.

Index

A

Advanced Plasma Power (APP), 84, 94
Arc discharge, 16–17
    atmospheric pressure
        air plasma composition of, 26
        argon plasma composition of, 25
    processes and properties, 20–21
        anode region, 23–24
        arc column, 24–26
        cathode region, 21–23
Arc plasma generators; *see also* Arc discharge
    alternative design, 17
    combination of, gas and water-stabilised, 18–19
    design goal, 16
    position stability, 16
    schematic arrangement, 17
    schematic configuration, 18–19
    thermal plasma production, 16
    torch characteristics
        analytic model, 27
        arc channel model, 27
        arc column, 27–29
        plasma enthalpy, 29–31
        steam plasma, 30–31
        torch efficiency, 29–30
    transferred and non-transferred plasma torches, 17
    transferred arc, 20
    tubular electrodes, 17
    water-stabilised plasma torch, 18

B

Biomass, 65–67, 87

C

Classical plasma, 10–11
Collisional plasma, 9–10
Combustion, 35, 51–56, 65–66, 88

D

DC technology, 91

E

Economical aspects
    advantages of, thermal plasma, 95
    electricity consumption, 96
    feasibility and efficiency, 95
    tipping fees, 95
    WTE tool/business model, 96–98
Electrical neutrality, 6
Electric field intensity, 29
Energos technology, 37
Energy balance, 27–28
    advantage, 69–70
    dependence of, plasma enthalpy, 62–63
    efficiency, 63
    heat combustion of, syngas, 65–66
    mass and heat balances, dry wood gasification, 67
    mass ratios, 69–70
    ratio of, LHV syngas, 68–69
    reactor volume, 64–65
    torch exit nozzle, 62
Environmental aspects, 98–99
Europlasma, 91, 94

G

Gasification, 51–56

H

Hazardous materials, 88–89

I

Incinerators, 34–35

L

Large-scale waste gasification
   concept demonstration, 36–37
   lower efficiency of, electricity generation, 36
   refuse-derived fuel (RDF), 36
Local thermodynamic equilibrium (LTE), 24–25

M

Maxwell velocity distribution, 1
Microwave plasma discharges, 15–16
MSW, *see* Municipal solid waste
Municipal sewage sludge, 89
Municipal solid waste (MSW), 84–87

N

Non-plasma systems, 45–47
Non-thermal plasmas, 12

P

Phoenix Solutions, 94–95
Plasma; *see also individual entries*
   classification of, 7
      classical plasma, 10–11
      collisional plasma, 9–10
      log n *vs.* log T representation, 8
      quantum/degenerate plasma, 7–9
      relativistic plasma, 7
   definition of, 1
   electrical neutrality, 6–7
   fourth state of matter
      characteristics, 4
      history, 3–4
      thermonuclear fusion, 4
   generation methods, 12–13
   mathematical descriptions, 11
   states of matter, comparison, 5–6
   temperature of, 1–3
   thermal *vs.* non-thermal, 11
Plasma-assisted pyrolysis, 51–56
Plasma gasification, *see individual entries*
Plasma-material heat transfer and process rate
   convection and radiation, 56
   dependence of, thermal conductivity, 56–57
   gasification rate of, wood particles, 60
   material particle, plasma flow, 58
   rate of, plasma gasification, 57–58
   surface temperature of, wood particles, 60
   volatilisation, 57–58
   volume ratio of, wood particles, 61
Plasma processing characteristics
   composition control, 46–47
   gasification of, organic substances, 45
   production of, syngas, 45–46
   temperature control, 47
Plasma reactor processes
   electric energy transformation, 47
   exothermic reactions, 49–50
   gaseous reaction, 47–48
   gasification of, organic materials, 49
   principal scheme, 48
   role of, 48
Plasma torches; *see also* Arc plasma generators
   applications, 20
   description, 16
   schematic arrangement, 17
   schematic configuration, 18–19
   transferred arc, 20
Product applications
   biomass, 87
   economical aspects
      advantages of, thermal plasma, 95
      electricity consumption, 96
      feasibility and efficiency, 95
      tipping fees, 95
      WTE tool/business model, 96–98
   environmental aspects, 98–99
   hazardous materials, 88–89
   municipal solid waste (MSW), 85–87

plasma gasification plants, industrial scale, 89–91
   Europlasma, 94
   Phoenix Solutions, 94–95
   Tetronics, 94
   Westinghouse Plasma Corporation (WPC), 92–94
single *vs.* two-stage plasma gasification, 84–85
sludge, 89–90
thermal plasmas advantages, 82–84
thermal plasmas technology, 81–82
Pyrolysis, 38–40, 51; *see also individual entries*

Q

Quantum/degenerate plasma, 7–9

R

Radio frequency (RF) discharges, 15–16
Refuse-derived fuel (RDF), 36, 84
Relativistic plasma, 7

S

Sludge, 89–90
Smaller scale waste gasification
   Energos technology, 37
   energy recovery, 38
   minimum emissions, 37

T

Tetronics, 91, 94
Thermal plasmas, 15, 39
   applications of, 82
   generation of (*see* Arc plasma generators)
   vs. non-thermal plasmas, 12
   waste treatment (*see* Waste treatment, thermal plasma)
Thermal *vs.* non-thermal plasma, 11
Thermochemical technology, waste treatment
   electricity/combined heat and power generation, 33

future applications, 38–40
gasification, 33–34
large-scale waste gasification
   concept demonstration, 36–37
   lower efficiency of, electricity generation, 36
   refuse-derived fuel (RDF), 36
smaller scale waste gasification
   Energos technology, 37
   energy recovery, 38
   minimum emissions, 37
waste incineration
   electrical efficiency, 35
   electricity generation, 34–35
Thermonuclear fusion, 4
Toxic organic waste, 88
Transition temperatures *vs.* transition energies, 5

V

Volatilisation, 47–51, 64–65

W

Waste incineration, 36
   electrical efficiency, 35
   electricity generation, 34–35
Waste treatment, thermal plasma
   combustion, 51–56
   energy balance
      advantage, 69–70
      dependence of, plasma enthalpy, 62–63
      efficiency, 63
      heat combustion of, syngas, 65–66
      mass and heat balances, dry wood gasification, 67
      mass ratios, 69–70
      ratio of, LHV syngas, 68–69
      reactor volume, 64–65
      torch exit nozzle, 62
   gasification, 51–56
   organics steam plasma, gasification
      energy balance of, wood gasification, 76
      input reagents, syngas composition and heating values, 72–73

mass and energy balances, 76
measured composition of, syngas, 75
oil gasification processes, 75
oxidising media, 71
steam and CO_2 plasmas, 70–71
thermodynamic equilibrium composition, 74
treated materials, 71
plasma-assisted pyrolysis, 51–56
plasma-material heat transfer and process rate
convection and radiation, 56
dependence of, thermal conductivity, 56–57
gasification rate of, wood particles, 60
material particle, plasma flow, 58
rate of, plasma gasification, 57–58
surface temperature of, wood particles, 60
volatilisation, 57–58
volume ratio of, wood particles, 61
processing characteristics
composition control, 46–47
gasification of, organic substances, 45
production of, syngas, 45–46
temperature control, 47
reactor processes
electric energy transformation, 47
exothermic reactions, 49–50
gaseous reaction, 47–48
gasification of, organic materials, 49
principal scheme, 48
role of, 48
Waste treatment, thermochemical process, *see* Thermochemical technology, waste treatment
Westinghouse Plasma Corporation (WPC), 92–94

Printed in the United States
by Baker & Taylor Publisher Services

Printed in the United States
by Baker & Taylor Publisher Services